BUSINESS/SCIENCE/TECHNOLOGY DIVISION
CHICAGO PUBLIC LIBRARY
400 SOUTH STATE STREET
CHICAGO, IL 60605

OCR →

THE CHEMISTRY OF
DIAMOND-LIKE SEMICONDUCTORS

The Chemistry of Diamond-like Semiconductors

N. A. GORYUNOVA

edited by
J. C. ANDERSON

translated by
SCRIPTA TECHNICA, INC.

THE M.I.T. PRESS
Massachusetts Institute of Technology
Cambridge, Massachusetts

First published (in Russian) 1963

Published in U.S.A. (in English translation) 1965 by
Massachusetts Institute of Technology Press,
Cambridge, Massachusetts

© Chapman and Hall Ltd, London, England, 1965

Library of Congress Catalog Card No. 65-24413

Ref
QD
475
G6313a

Printed in Great Britain by
Fletcher & Son Ltd, Norwich

CONTENTS

		page	
	Preface		vii
	Introduction		ix
1.	Formation of Diamond-like Substances		1
2.	Physicochemical and Electrical Properties of Diamond-like Semiconductors		58
	A. Elemental Semiconductors		58
	B. Binary Compounds		83
	C. Isovalent Solid Solutions		124
	D. Ternary and More Complete Heterovalent Phases		139
	E. Defect Tetrahedral Phases		152
	F. Excess Tetrahedral Phases		176
3.	Principles Governing Variations in Properties of Diamond-like Semiconductors		193
	Conclusion		214
	References		219
	General Index		237
	Formula Index		241

PREFACE

This monograph deals with the fundamental problems of the chemistry of diamond-like semiconductors. These problems are closely related to the general theory of solid state physics, the theory of semiconductors, crystal chemistry and physicochemical analysis. All this material has been far from exhausted; in the majority of cases, problems bordering on scientific fields other than those listed have had to be dismissed with only cursory remarks. The outlines of the chemistry of diamond-like semiconductors have been formulated, by the author of this book and her co-workers, in the light of the developments in this field from 1950 onwards.

The fundamental ideas underlying the chemical investigations of semiconductors, have been formulated, and material scattered throughout hundreds of literature sources has been collected together, and an attempt has been made to present it in a systematic form, capable of providing a direction for future investigations, all in a comparatively small volume.

Not all sections of the book are treated by the author with equal thoroughness. Thus compounds of type A^2B^6, which have been known to chemists for a long time, are described more briefly than complex compounds. Excess tetrahedral phases, which have been insufficiently studied from the standpoint of semiconducting properties, but have been referred to frequently in the literature, are presented only in the form of tables with the corresponding references. The problems of the alloying and interaction of admixtures are given little attention due to lack of space. The magnetochemistry of diamond-like semiconductors, an important but specialized and little-developed field, has not been treated at all.

In such a rapidly developing field as the chemistry of semiconductors, where research work is being carried on in several countries (in many cases along identical lines) and new data are being published in large quantities, the monograph should reflect the present position.

For this reason, the book was written, taking fully into consideration new experimental data as well as general and theoretical aspects which have come to light most recently. The lectures in semiconductor chemistry which I delivered in the chemistry department of Leningrad University in 1958-1961 and my dissertation of 1958 have been used only partially.

The difficulties inherent in this task may have led to certain problems and vague points in presentation, and also to deficiencies in the selection and distribution of the material. I shall therefore appreciate any comments which the reader may offer.

I express my sincere and profound thanks to my co-workers at the semiconductor laboratory of the A. F. Ioffe Physicotechnical Institute of the Academy of Sciences of the U.S.S.R., particularly N. K. Takhtareva, A. A. Vaipolin, E. V. Tsvetkova, V. I. Sokolova, L. V. Kradinova, E. Yu. Lubenskaya and I. I. Tychina, for their assistance in the writing of the manuscript and its preparation for press.

The author

INTRODUCTION

THE chemistry of semiconductors came into being as a new science when it became clear that physical phenomena in semiconductors were based on the chemical properties of solids, i.e., the positions of the elements in the periodic system and, hence, the character of their electronic interactions and the crystal structure.

During the first stage of development of the science of semiconductors, chemistry played a subordinate part in this field and was chiefly preparatory in character. Only certain eminent scientists, including A. F. Ioffe, understood the significance of the chemical nature and properties of matter in the theory of the solid state and in its practical application. Thus, the notion of the importance of short-range order to the electrical properties of solids is due to A. F. Ioffe, who discussed this concept, including both the chemical composition and the geometry of the arrangement of atoms in space[1].

As might have been expected from the very start of the investigations, the chemical nature of semiconducting substances, particularly diamond-like substances, is determined by the presence of a pronounced covalent bond between the atoms and by the characteristics of this type of electronic interaction. However, simple inorganic substances and compounds with covalent bonds were studied relatively little in the early days.

It is well known that the inorganic chemistry of the 1930's developed largely as the chemistry of acids, bases and salts. The theory of chemical structure originated by A. M. Butlerov, based on the concept of rigid directional covalent bonds, was applied primarily to organic compounds, since the latter exhibit properties determined by the structure of matter, and not only by its composition, to a much greater extent than do salts.

INTRODUCTION

Until comparatively recently, the only area of inorganic chemistry which involved the development of concepts of covalent bonds was the chemistry of complex compounds. Finally, toward the end of the thirties, R. L. Myuller put forward the idea of the participation of covalent bonds (atomic-valent bonds, according to the author's terminology) in inorganic glasses[2]; this idea was developed further in investigations of refractory and chalcogenide glasses.

Only in the last few years have the structure and properties of an enormous number of crystalline chemical compounds which are semiconductors, been qualitatively explained on the basis of the concept of directed valences.

In semiconductors, particularly those of the diamond-like type, owing to a high symmetry and the completeness of the electron clouds constituting the bonds between the atoms, both the physical and the chemical properties are clearly determined by what A. M. Butlerov called chemical structure.

It became evident that the inorganic chemist's basic problems concerning the methods of preparation of substances and their interaction with other substances assumed a special significance when applied to binary diamond-like semiconductors. The latter have, as a rule, very high melting points compared to those of the parent substances, and a very narrow region of homogeneity, practically limited by the stoichiometric ratio, and they are insoluble in water.

The covalent bond, which determines these properties, is also responsible for the fact that the actual manifestation of the periodic law in the properties of the diamond-like semiconductors has its specific peculiarities. These peculiarities, as objects of physico-chemical analysis, determine the modes of approach to their study, in contrast to salt or metallic systems.

It is important to note that from the very start of investigations into the chemistry of diamond-like semiconductors, these studies were built on crystallochemical concepts (e.g., the concepts of a crystallochemical group, etc.).

An important factor promoting a fruitful mutual influence between semiconductor physics and chemistry was the possibility of evaluating the nature of the chemical bonds in solids on the basis of investigations of the electric properties (electric conductivity in the region of intrinsic conduction, width of the forbidden band, mechanism of scattering of the current carriers, etc.).

Methods of other allied fields, e.g., metallurgy, were also applied to further the chemistry of semiconductors. To obtain the substances in a form suitable for studying their electric properties, it was found necessary to use the methods of zone refining, zone levelling, etc.

The beginning of the introduction of chemical concepts into the science of semiconductors dates back to the early post-war years, when a relationship was established between the electric parameters of germanium, silicon, gray tin, their structure, and the type of chemical interaction between the atoms. Then physicists and chemists in different countries began almost simultaneously to study the binary analogues of elements of group IV, in addition to the chemical properties of various semiconductors. Among these elements were found substances with an extremely high carrier mobility, a parameter of prime importance for a great many practical applications. This attracted the attention of researchers, and the number of studies in this field began to increase rapidly, making it possible to treat from a single point of view a large group of substances similar in structure and in the type of chemical bonding. It is very significant that such treatment led to correct predictions of the practically important properties of substances in this group which were still unknown at that time.

Thus the experimental and theoretical investigations of the group of diamond-like semiconductors were the first contribution of the new science of semiconductor chemistry to physics and engineering. Substances of the diamond crystallochemical group have extensive potential applications as materials for semiconductor devices. Many of them have already found application in radio electronics. This was due to the special properties of the tetrahedral covalent bonds characteristic of the entire crystallochemical group of diamond. Bonds of this type create the best conditions for electron transfer, a phenomenon lying at the basis of the operation of semiconductor devices.

Compounds of the type A^3B^5, the closest analogues of the semiconductor elements of group IV of the periodic system, have found application in semiconductor engineering. Binary semiconductors of the type A^3B^5 were found to possess a combination of main physicochemical and electrical parameters which was different from that of diamond, silicon, germanium, gray tin and solid solutions based on them. For instance the intrinsic width of the forbidden band, the mobility of the majority carriers and the melting point in the diamond-gray tin group are such that a width of the forbidden band of more than 1 eV is necessarily associated with majority carrier mobilities of less than 2000 cm^2/V sec and melting points at temperatures above 1200°C. In compounds of type A^3B^5, such as gallium arsenide, it is possible to obtain 115 times the width of the forbidden band and twice the majority carrier mobility for the same melting point of the material.

A still greater possibility of variation of these parameters is provided by substitutional solid solutions based on compounds of type A^3B^5. Thus, for example, on the basis of the solid solutions mGaSb:$(1-m)$InSb first studied in the Soviet Union it is possible to obtain semiconductor materials with a width of the forbidden band three times as large as that of gray tin and with an electronic mobility exceeding that of germanium by one order of magnitude.

However, practical requirements outstrip the actual possibilities of the science of semiconductor materials. New substances with new qualities and combinations of properties are constantly needed. It is precisely these materials which can serve as the basis for fundamentally new devices, just as, some years ago, the new devices were based on germanium and several years later on indum antimonide. For this reason, the search for new semiconductor materials continues at a rapid pace throughout the world.

In the field of diamond-like semiconductors, one of the trends of this research is concerned with the elucidation of the possibility of preparing practically important materials on the basis of complex, multicomponent substances of the diamond-zinc blended crystallochemical group.

A second, no less important, trend is the extension of the crystallochemical group to substances with similar and related structures such as those of the fluorite type, which also have tetrahedral bonds with the aim of finding substances possessing optimum properties for various applications.

In the entire field of semiconductor chemistry, particularly the chemistry of diamond-like semiconductors, the fundamental problem determining the basic trends of the research is that of correlating the electric (semiconducting) properties of substances with the chemical composition, the crystal structure, and the nature of the bonding.

The purpose in studying this problem is the elucidation of the physicochemical and crystallochemical principles of the creation of semiconducting substances with predetermined properties. An attempt will be made in this book to show to what extent this problem has been solved for the group of diamond-like semiconductors.

CHAPTER 1

FORMATION OF DIAMOND-LIKE SUBSTANCES

1. Elements Forming Tetrahedral Phases

THE periodic variation of the properties of elements with increasing atomic weight, discovered by D. I. Mendeleev, was used by the latter as the basis for predicting the properties of elements not yet studied. These properties were defined as being intermediate between those of the neighbouring elements on the left and right as well as top and bottom in the periodic system. A still greater similarity may be observed between an element and its neighbours on the left and right in the periodic system when the latter form a chemical compound with one another.

This characteristic is manifested most distinctly by elements in group IVb. A far-reaching similarity is seen in this case between these elements and the chemical compounds formed by the neighbouring elements. Thus, many binary compounds of elements of groups III and Vb, embraced by the general formula A^3B^5, have been found to be in many respects (such as structure and electric properties) similar to the elements of group IVb.

V. M. Goldschmidt[3] was the first to point out the existence of binary compounds formed by elements equidistant from group IV of the periodic system which have the wurtzite and sphalerite structures, similar to the structure of diamond (Fig. 1). The diamond, sphalerite and wurtzite structures, characteristic of the elements and binary compounds of this group, are characterized by a coordination number of 4. By analogy with diamond, the electron density in the tetrahedral directions is considered to be maximum.

Goldschmidt synthesized many of these compounds, having determined their lattice constants and the Mohs hardness, and found that in horizontal series such as Ge – GaAs – ZnSe – CuBr or grey Sn – InSb – CdTe – AgI (these series were later called isoelectronic), the lattice spacings of the compounds and those of the corresponding substances of the fourth group were very close, and that the hardness changed regularly with the "strain of the binding forces", i.e., with the change in the valence of the atoms constituting the compounds.

The surprising closeness of the lattice spacing of the compounds of the same series was attributed by Goldschmidt to the non-ionic interaction of the components of these compounds (if the interaction were ionic, the interatomic distances would decrease with an increase in the valence of the atoms, something that is not observed in the isoelectronic series).

These interesting results are due to the fact that group IV is the axis of symmetry of the periodic system. However, this fact led to results which were still more important from the practical point

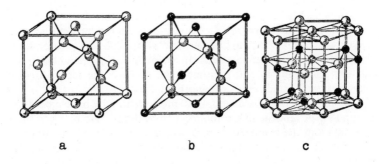

Fig. 1 *Structures of (a) diamond (b) sphalerite and (c) wurtzite*

of view. A large amount of experimental work has established that the physical and physicochemical properties of these substances are shared qualitatively to a very considerable extent. It was found possible, therefore, to incorporate in the single crystallochemical group of semiconductors the compounds A^3B^5, which are analogues of group IV, as well as the compounds A^2B^6 and A^1B^7, which have the same structure and properties but are more remotely reminiscent of the properties of diamond and of members of the

FORMATION OF DIAMOND-LIKE SUBSTANCES

same sub-group[4,5].

Within this group, which includes elements and binary compounds, and also, as will be apparent below, more complex substances, the properties vary regularly according to the positions which the elements constituting the substances occupy in the periodic system. This served as the basis for predicting the properties of substances still unknown.

Table 1 shows the crystallochemical group of diamond-like semiconductors (elemental and binary) arranged in isoelectronic series; all the members of this group except boron antimonide are obtained in the pure state. Column A^4 contains, in addition to diamond, silicon, germanium and gray tin, hypothetical alloys whose analogues are the corresponding binary compounds.

Table 1 Isoelectronic series of the crystallochemical group of diamond-like semiconductors

A^4	A^3B^5	A^2B^6	A^1B^7	A^4	A^3B^5	A^2B^6	A^1B^7
2C	BN	BeO		αSn + C	InN	ZnSe	CuBr
C + Si	BP	BeS		2Ge	GaAs	MgTe	
Si + C	AlN			Si + αSn	AlSb	CdS	
2Si	AlP			αSn + Si	InP		
C + Ge	BAs	BeSe		Ge + αSn	GaSb	ZnTe	CuI
Ge + C	GaN	ZnO		αSn + Ge	InAs	CdSe	
Si + Ge	AlAs			Pb + Si		HgS	
Ge + Si	GaP	ZnS	CuCl	2αSn	InSb	CdTe	AgI
C + αSn	BSb?	BeTe		Pb + Ge		HgSe	
				Pb + αSn		HgTe	

As can be inferred from an inspection of Table 1, the formation of binary compounds with a tetrahedral structural arrangement of the atoms* is most characteristic of elements of the middle portion

---
* Here and elsewhere in the text we use the expressions "tetrahedral structural arrangement of the atoms" and "fourfold coordination" to mean an atomic coordination where each atom is surrounded by atoms of a different kind. In this case, the terms do not cover structures where the central atom is surrounded by a tetrahedral arrangement which includes atoms of the same elements as well as atoms of other elements.

of the periodic system, those with filled 18-electron shells.

A more graphic representation of the participation of elements in simple and binary substances with a tetrahedral structural arrangement of the atoms is given in Table 2, which shows a part of the periodic system, leaving out the even series of the long periods*, where the bold type indicates the elements which form binary compounds with a tetrahedral structural arrangement of the atoms with

Table 2 Elements participating in the formation of tetrahedral structures in binary compounds

Li	Be	B	C	N	O	F
Na	Mg	Al	Si	P	S	Cl
Cu	Zn	Ga	Ge	As	Se	Br
Ag	Cd	In	Sn	Sb	Te	I
Au	Hg	Tl	Pb	Bi	Po	At

at least one of the elements equidistant from group IV. The dotted squares enclose elements giving only the wurtzite structure in analogous compounds.

The position of the group of binary compounds with the structure of sphalerite and wurtzite among other inorganic compounds of the same formula AB is shown in Table 3.

It should be noted that the data of Table 3 do not pertain to the structure of the substances in thin-film form. It is well known that in thin films, a considerably greater number of compounds than that shown in Table 2 can crystallize in the form of the two modifica-

---
* Of the elements in the even series of the long periods, only manganese forms two compounds (the sulphide and the selenide), which crystallize in the structure of zinc blende.

tions, sphalerite and wurtzite. Table 3 demonstrates the character of morphotropic transitions at the boundaries of a crystallochemical group.

Table 3 Structures of compounds formed by elements of the short series of the periodic table

	A¹B⁷					A²B⁶					A³B⁵				
Elements	F	Cl	Br	I	*Elements*	O	S	Se	Te	*Elements*	N	P	As	Sb	Bi
Li	NaCl	NaCl	NaCl	NaCl	Be	w	s	s	s	B	s	s	s	?	—
Na	NaCl	NaCl	NaCl	NaCl	Mg	NaCl	NaCl	NaCl	w	Al	w	s	s	s	—
Cu	?	s	s	s	Zn	w	w, s	w, s	w, s	Ga	w	s	s	s	—
Ag	NaCl	NaCl	NaCl	w, s	Cd	NaCl	w, s	w, s	s	In	w	s	s	s	CsCl
Au	—	?	?	?	Hg	HgO	s	s	s	Tl	—	—	—	CsC	CsCl*

Arbitrary designations of structures: s – sphalerite type; w – wurtzite type; CsCl – distorted structure of CsCl type; dash – compound does not exist; question mark – structure of the compound unknown

The principles of the formation of compounds of the diamond crystallochemical group do not obey the law of Magnus, which restricts the formation of structures of fourfold coordination within the following limits of the ratio of the ionic radii K:

$$0.41 \gg K \geqslant 0.225. \quad K = \frac{r_\text{к}}{r_\text{a}},$$

where r_c is the cation radius and r_a the anion radius.

As can be easily calculated, of the 34 compounds with the structures of sphalerite and wurtzite, only 14 correspond to radius ratios which are in the range of values specified by the law of Magnus. The non-adherence to the latter emphasizes that one bond type is specific to substances of this group, and the approach based on the concept of the closest packing of ion spheres[3] is inapplicable. Using Magnus' ideas, Goldschmidt supplemented them with considerations of polarization, but did not obtain a satisfactory agreement with the experimental data.

Despite the inadequacy of the arguments involving polarization, they are still being used to explain the causes of the formation of structures with a tetrahedral arrangement of atoms. Most such explanations are based on the fact that all the elements of the main subgroups of the periodic system which produce cations with 18-electron shells form compounds with the structure of sphalerite or wurtzite, since these cations have the strongest polarizing effect and polarize with relative ease themselves.

The formation of tetrahedral structures is often attributed to the tendency[6,7] of cations formed from atoms with outer *dsp* electrons to form lattices with a coordination number of less than 6, and in particular, a tetrahedral configuration; this amounts essentially to the preceding statement.

As can be seen from Table 2, this is not confirmed by experimental data: beryllium, boron, aluminium and magnesium are elements which do not have *d* electrons, and which nevertheless form compounds with the structure of wurtzite and sphalerite.

In its most general form, the question of the formation of directional valence bonds was discussed by Kimball[8]. Using group theory, he found all the possible stable electron configurations, particularly for the coordination number 4. However, the general form of this solution is insufficient for elucidating the problem of the concrete participation of specific elements in tetrahedral structures.

Mooser and Pearson[9] suggested the use of the mean principal quantum number of the valence shells of atoms forming a compound, and the difference in the electronegativity of these atoms as a measure of the metallic and ionic character respectively. On the basis of these parameters, which in their view are the criteria for the directionality of the bonds, Mooser and Pearson divided the compounds of types AX, AX_2, etc. into groups with various coordination configurations and structures.

Fig. 2 shows a graph taken from the above work, giving the classification of normal valence compounds of the type AX formed by cations of subgroup A and showing also the separation of the structures of zinc blende (*B*-3), wurtzite (*B*-4), rock salt (*B*-1) and caesium chloride (*B*-2). Similar graphs exist for compounds of other types.

In the view of Mooser and Pearson, the "striking character of this separation suggests that these two parameters, also related to the magnitude of the forbidden band of semiconductors, should be an effective means of evaluating the activation energy of semiconducting compounds",[10, p. 251]. We shall show below that the same separation of structures can also be obtained by using entirely different parameters. There is therefore no reason to ascribe such exceptional importance to the Mooser and Pearson parameters.

Attempts have also been made in the literature to present the formation of covalent compounds from other points of view. In 1946, B. V. Nekrasov proposed that the polarity of valence bonds be estimated by means of the electron affinity constants. These are the energies involved in the addition to the isolated atom[11] of the first electron beginning the formation of the next valence-electron shell, i.e. the ionization potentials of the first electron for univalent cations, of the second for divalent cations, etc.

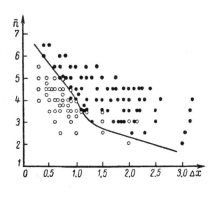

○ — tetrahedral coordination
● — octahedral coordination

Fig. 2 *Normal valence compounds with structures of sphalerite, wurtzite, chalcopyrite and rock salt on a graph of principal quantum number n versus electronegativity difference* $\Delta\chi$[9]

Later, Ahrens[12] proposed the use of the same quantities for the comparative evaluation of electrostatic fields. Morris and Ahrens[13] showed that there exist limiting values of the nth ionization potential of elements which permit the separation of covalent and ionic compounds. The authors consider the concept of electronegativity less suitable for the treatment of bonds in crystals, since electronegativity pertains to conditions of stable equilibrium, while the method of these authors is related to the process of formation of a crystal and makes it possible to predict its properties.

The authors show for each individual group that, as the ionization potential increases, a transition from a typically ionic to a covalent bond is observed. However, since they consider only the ionization potentials of the first three groups, they have to use the polarizability, a quantity that can be determined only very roughly, when this relation breaks down. They also have to use ionic radii in this treatment, and establish relationships for "cations of large radii" and "cations of small radii" separately; this, however, does not help to form a structural picture which would permit the delineation of the limits to the formation of ionic and covalent compounds.

In our view, the most effective means of developing a theory of formation of crystal structures (which does not as yet exist) could be the elaboration of an empirical chemical model, reflecting the most important stages of formation of any given structures, with a definite character relating to the electronic interaction of the constituent atoms.

As follows from experiments involving the determination of electron densities, the basis of the structure of substances with a tetrahedral arrangement of atoms is the electronic lattice formed by the tetrahedrally arranged "bridges" composed of paired valence electrons. The "bridges" are more or less diffuse, depending upon the atomic weight of the elements comprising the compounds (or upon their principal quantum number) and are more or less symmetrical (with respect to the distribution of the electron density along the "bridge") depending upon the differences in the electron affinity of the units occupying the lattice points. Such units are atoms deprived of their valence electrons. Thus, the electronic lattice is the result of the interaction of atoms whose ionic cores are located at the lattice points. The bulk properties of substances in a given crystallochemical group are determined by the parameters of the electronic lattice and those of the kernels. Details of the structure of the electronic lattice can be determined only from indirect data; only the quantum numbers of the atoms are known exactly. As far as the conic cores are concerned, they can be characterized by the ionization potentials.

These parameters may provide information on whether the elements can form compounds with certain types of chemical bonds. From this standpoint, the initial assumptions of Nekrasov and Ahrens seem to us to be fully acceptable. However, as will be seen further, they also need to be developed.

The formation of substances with ionic and metallic bonds is energetically advantageous when the energies of detachment of the valence electrons of atoms (or ions in the case of an ionic bond) are small.

Indeed, it is known that metallic elements and the elements present in metallic compounds have low ionization potentials. The elements of the right-hand side of the periodic table with high ionization potentials of the valence electrons form compounds with one another whose interatomic bonds are typically covalent. An ionic bond, however, is known to require a more complete transfer of electrons from the cation to the anion than in the case of a covalent bond; i.e., it is more advantageous energetically at small values of the ionization energy of the valence electrons of one of the atoms. Hence it is obvious that in trying to assess the possibility of the formation of any given structures, it is desirable to use the electron affinity constants N proposed by Nekrasov. Then the formation of tetrahedral structures becomes possible if the electron affinity constant is greater than a certain value for each group of elements; the formation of compounds with ionic or metallic bonds is then not energetically advantageous.

A further step can be made along this path if, as suggested by Goldschmidt [14], we introduce specific electron affinity constants, i.e., if we divide the electron affinity constants by the charge on the ionic core

$$N' = \frac{N}{n}.$$

We thus introduce a characteristic of the force field of the ionic core per unit charge.

Table 4 gives the values of specific electron affinity constants for the most common elements, calculated with the assumption that the latter achieve a normal valence. An exception is made only for elements of the ninth series (Tl, Pb, Bi), where, because of the particular stability of the $6s^2$ subshell, the valence which is usual for them is reduced by two units. In addition, beginning with group IV, the specific electron affinity constants for elements of the inserted decades have been calculated from the second ionization potential. In the case of compounds of these elements with a valence different from two, the values of the constants should be changed correspondingly.

Table 4 Specific electron affinity constants of the elements

series number	Elements	N',v	series number	Elements	N',v	series number	Elements	N',v
1	H	13.54	24	Cr	8.20	46	Pd	9.92
3	Li	5.37	25	Mn	7.25	47	Ag	7.58
4	Be	9.06	26	Fe	7.95	48	Cd	8.40
5	B	12.57	27	Co	8.73	49	In	9.28
6	C	16.02	28	Ni	9.44	50	Sn	10.18
7	N	19.46	29	Cu	7.67	51	Sb	11.10
8	O	22.88	30	Zn	9.02	52	Te	11.98
9	F	26.31	31	Ga	10.22	53	I	12.9
11	Na	5.09	32	Ge	11.38	55	Cs	3.86
12	Mg	7.55	33	As	12.52	56	Ba	4.98
13	Al	9.45	34	Se	13.65	57	La	6.37
14	Si	11.21	35	Br	14.78	75	Re	6.58
15	P	12.92	37	Rb	4.19	78	Pt	8.68
16	S	14.58	38	Sr	5.43	79	Au	9.20
17	Cl	16.26	39	Y	6.82	80	Hg	9.28
19	K	4.32	40	Zr	6.98	81	Tl	6.08
20	Ca	5.93	41	Nb	6.74	82	Pb	7.45
21	Sc	8.73	42	Mo	7.58	83	Bi	8.37
22	Ti	6.80	44	Ru	8.19			
23	V	7.56	45	Rh	9.03			

It is obvious from a comparison of Tables 4 and 2 that the condition of formation of tetrahedral structures for elements of all groups may be expressed by the single inequality

$$N' = \frac{N}{n} > 7\cdot 25 \text{ V},$$

i.e., the formation of compounds of type AB with a tetrahedral structural arrangement of atoms requires that the specific electron affinity constant of each of the constituting elements be greater than 7·25 V.

If this inequality is satisfied, a further evaluation of the possibility of formation of tetrahedral structures requires an examination of the specific electron affinity constants of both components together. Fig. 3 shows a plot of these data for the same compounds as those given by Pearson in Fig. 2.

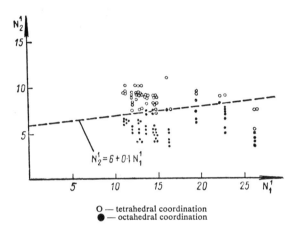

O — tetrahedral coordination
● — octahedral coordination

Fig. 3 *Separation of structures of type AB compounds as a function of the specific electron affinity constants N^1_1 and N^1_2 of the constituent*

It is evident from a comparison of Figs. 3 and 2 that the specific electron affinity constants lead to the same separation of compounds according to structure as that obtained by means of Pearson's parameters, but are undeniably better suited for experimental determinations than the kind of rough numerical estimate which is provided by the difference in electronegativities. In Fig. 3, the boundary between the regions occupied by the tetrahedral and octahedral structures may be approximately expressed by the straight line

$$N'_2 = 6 + 0 \cdot 1 N'_1$$

whose equation relates the values of the specific constants of the components and enables one to evaluate the crystal structure of the substance. A similar separation of structures as a function of the specific electron affinity constants can also be obtained for other types of compounds.

An evaluation of the possibility of formation of any given structures may also utilize such parameters as, for example, the total group potentials. All this points to the fact that this problem has not yet been finally resolved. Nevertheless, such considerations will permit definite conclusions and predictions with regard to the formation of any given structures, this being very important in the search for new semiconducting materials.

2. Complex Tetrahedral Phases

Although all of the above pertained to binary tetrahedral phases, more complex phases are usually formed by those elements whose energetic atomic characteristics satisfy the above-mentioned rule.

The problem of the participation of any given elements in the formation of a tetrahedral structure may be considered to have been clarified to a certain extent. However, the problem of the formation of tetrahedral phases more complex than the binary ones involves consideration of the possibility of the formation of tetrahedral phases by elements belonging to different groups of the periodic system which are present in various proportions.

First of all, it is necessary to emphasize the fact that from here on, multicomponent tetrahedral phases should be taken to mean phases whose components belong to different groups of the periodic system, in contrast to the multicomponent phases formed by simpler components via isovalent substitution (see Part C, Chapter 2).

As far back as the 1920's, Grimm and Sommerfeld[15], using the experimental data of Goldschmidt, formulated a rule for the formation of binary compounds of type AB with a tetrahedral coordination of the atoms in the structure. This rule involved two conditions of formation: (1) the elements must belong to groups of the periodic system which are equidistant from group IV, and (2) the average number of valence electrons per atom must be four. Soon, however, binary tetrahedral phases with a different proportion of atoms ($A_2^3 B_3^6$) were discovered, as well as ternary phases, for example, $A^1 B^3 C_2^6$, $A^2 B^4 C_2^5$, $A^1 B_2^4 C_3^5$, $A_2^1 B^4 C_3^6$ and $A_3^1 B^5 C_4^6$ (see Part D, Chapter 2). All these substances were found to have the properties of semiconductors.

The phases having the formula $A_2^3 B_3^6$ were classified in the group of defect semiconductors of zinc blende structure on the basis of X-ray studies, since a proportion of the lattice points in these phases remain vacant. As far as ternary tetrahedral phases are concerned, their structure and properties, which are close to those of binary tetrahedral phases, led to their unification in a single group of analogues of elements of group IV, whose formation obeys laws common to the entire group of analogues, regardless of the number of components.

This gave rise to the problem of expanding the Grimm-Sommerfeld rule. Following this course, Goodman used the idea of transverse substitution, which he got by treating the formation of compounds of type A^3B^5 as analogues of substances A^4 [16]. From a chemical point of view, the principle of heterovalent transverse substitution is not a new one.

The ideas of Fersman and Vernadskiy on the heterovalent substitution in minerals are widely known. An interesting example may be cited for the substitution of phosphorus atoms by an equal amount of sulphur and silicon in apatites[17]. Heterovalent substitutions within the confines of the zinc blende structure were also discovered a long time ago[4, 18]. These experiments were the beginning of an investigation of a large number of systems with a heterovalent substitution of atoms, even before the theoretically possible types of such systems had been found.

Using the method of transverse substitution, Goodman obtained theoretically ternary diamond-like phases; the existence of part of these phases was confirmed experimentally. However, the method of transverse substitution is not rigorous enough for more complex phases such as quaternary ones.

Thus, for example, the method of transverse substitution does not explain the fundamental difference in the width of the homogeneity region of binary and ternary phases on the one hand and more complex tetrahedral phases on the other.

If an attempt is made to derive certain obvious conditions of formation for tetrahedral phases from the available experimental material and then to apply them to the problem of the number and type of multicomponent tetrahedral phases, the solution of this problem turns out to be more rigorous and exhaustive[19]. It is known that the similarity in the properties of elements in the same group of the periodic table is due to the fact that the number of their valence electrons is the same. Therefore, when the question arises as to what is the composition of complex compounds made up of elements of different groups and possessing the same properties as simple substances, it is natural to assume that, on average, the number of valence electrons per atom in the complex substances should be the same as in the simple ones.

The tetrahedral arrangement of atoms in the structure of diamond and zinc blende requires an average number of valence electrons per atom equal to four. Hence, for complex phases of this group which are analogues of the elements in group IV, the following rule may be formulated: their composition should be such that the average number of valence electrons per atom is four. This rule immediately imposes severe limitations on the number of possible combinations of elements of different groups among which one should look for analogue phases of group IV. For instance, the number of combinations of two elements from seven different groups of the periodic table will be

$$C_m^n = \frac{m!}{n!\,(m-n)!} = \frac{7!}{2!\,(7-2)!} = 21.$$

This rule can of course be fulfilled for combinations in which one of the elements belongs to a group to the left of group IV, i.e., has a number of valence electrons of less than four, while the second element has more than four. Twelve combinations can thus be dropped from consideration. Shown below are binary compounds with four valence electrons per atom (9 combinations).

System	Formula	System	Formula
1—5	$A^1B_3^5$	2—7	$A_3^2B_2^7$
1—6	$A_2^1B_3^6$	3—5	A^3B^5
1—7	A^1B^7	3—6	$A_2^3B^6$
2—5	$A^2B_2^5$	3—7	$A_3^3B^7$
2—6	A^2B^6		

Of the above compounds, only three, viz., A^3B^5, A^2B^6, and A^1B^7 are analogues of elements of group IV. By comparing them with all the others, it is easy to discover the second rule which is satisfied by the binary phase analogues of group IV; namely, if the element of the smaller group is arbitrarily regarded as the cation and that of the larger group as the anion, an analogue phase is formed only if a normal valence is achieved; the valence is normal in the sense that the number of electrons given up by the "cation" to form tetrahedral bonds should be equal to the number of electrons needed to complete an octet in the "anion". It may be assumed that this rule, which holds for binary compounds, will also apply to more complex systems.

If the capital letters in the formula of a binary compound $A_{(1-x)}B_x$ are understood to mean not only the symbol of the element but also the number of its valence electrons, and if we always consider $A < B$, the rule for the achievement of normal valence will be mathematically written as

$$A(1-x) = (8-B)x,$$

and the rule formulated earlier according to which the number of valence electrons per atom should be equal to four will be written as

$$A(1-x) + Bx = 4.$$

These rules may be written in similar fashion for more complex systems as well. Using them, we shall now look for the analogue phases of group IV having a complex structure.

Ternary systems. The general formula of a ternary system will be $A_{(1-x-y)}B_xC_y$, where we shall assume that $A < B < C$. The number of possible combinations of three elements from seven different groups is

$$C_m^n = \frac{7!}{3!(7-3)!} = 35.$$

In addition, an intermediate element of the system may play the part of either a "cation", and the system is then "dicationic", or of an "anion" in a monocationic system*. This makes it necessary to consider all of the 35 possible systems in two variants, i.e., the total number of possible systems will be 70.

Let us consider, for example, the "dicationic" variant of ternary systems. Here the condition of normal valence will be written as

$$A(1-x-y) + Bx = (8-C)y,$$

---

* Since we are dealing with compounds having a pronounced covalent bond type, the terms "cation" and "anion" used here and below are given in quotation marks.

and the four-electron condition will be

$$A(1 - x - y) + Bx + Cy = 4.$$

Solving these two equations simultaneously, we find

$$y = 1/2;\ x = \frac{8 - A - C}{2(B - A)}.$$

Hence we conclude that the "dicationic" ternary analogue phases of group IV will be only those whose general formula is

$$A_{(1/2 - x)}B_x C_{1/2},$$

with the following restriction:

$$0 < x = \frac{8 - A - C}{2(B - A)} < 1/2.$$

It is now easy to see that of the 35 possible systems, only the five given below satisfy the inequality obtained:

Ternary "dicationic" phases

System	Formula	System	Formula
1—3—6	$A^1B^3C_2^6$	1—5—6	$A_3^1B^5C_4^6$
1—4—5	$A^1B_2^4C_3^5$	2—4—5	$A^2B^4C_2^5$
1—4—6	$A_2^1B^4C_3^6$		

Representatives of all these combinations have been identified and, from the standpoint of their properties may be considered to be analogues of elements of group IV.

Let us examine the second variant, the "monocationic" ternary systems. In this case, the condition of maximum valence will be written differently:

$$A(1 - x - y) = (8 - B)x + (8 - C)y,$$

and the four-electron condition will remain unchanged:

$$A(1 - x - y) + Bx + Cy = 4.$$

Simultaneous solution gives

$$x + y = 1/2; \quad y = \frac{8 - A - B}{2(C - B)}.$$

Hence, the formula of "monocationic" ternary analogue phases of group IV assumes the form

$$A_{1/2} B_{(1/2 - y)} C_y,$$

provided that

$$0 < y = \frac{8 - A - B}{2(C - B)} < 1/2.$$

Of the 35 possible systems, the following five satisfy this inequality:

Ternary "monocationic" phases

System	Formula	System	Formula
2—5—7	$A_2^2 B^5 C^7$	2—3—7	$A_4^2 B^3 C_3^7$
3—4—7	$A_3^3 B_2^4 C^7$	3—4—6	$A_2^3 B^4 C^6$
2—4—7	$A_3^2 B^4 C_2^7$		

These phases are mirror images of the above "dicationic" tetrahedral phases with respect to group IV of the periodic table. Attempts to obtain substances of such types have shown that almost all of them are unstable or crystallize in complex structures.

Hence one can conclude that among more complex systems of analogues as well, interest should centre primarily on those where the number of elements of the "cations" is greater than or equal to that of the "anions".

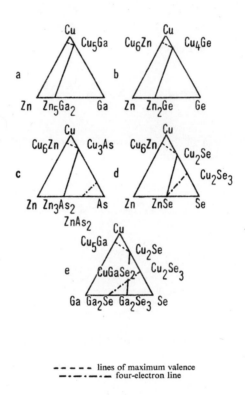

- - - - - lines of maximum valence
— · — · — · — four-electron line

Fig. 4 *Formation of ternary tetrahedral analogue phases of germanium*

What was stated above concerning the formation of ternary analogue phases of group IV can be illustrated by drawing triangular

FORMATION OF DIAMOND-LIKE SUBSTANCES

concentration diagrams (Fig. 4). As an example, we shall take the elements of the fourth period, which participate in the formation of tetrahedral bonds. As we know, according to the principles of physicochemical analysis, each point inside the triangle represents a combination of three elements taken in a certain proportion. If any of the stated conditions is applied, however, one obtains straight lines of pseudobinary sections, while the intersection of two such lines gives a point satisfying both conditions simultaneously. Then, depending upon the groups to which the elements under consideration belong, the following cases of arrangement of these straight lines are possible:

(*a*) The straight line corresponding to the four-electron condition is absent (Fig. 4*a*);

(*b*) The straight line is represented by one point at the vertex of the triangle (Fig. 4*b*). The analogue of group IV is the element of group IV itself;

(*c*) the straight line is present, but does not intersect with the lines of maximum valence inside the triangle (Fig. 4*c*). Here also the analogues of group IV are absent;

(*d*) the straight line intersects with one of the lines of maximum valence on the side of the triangle (Fig. 4*d*). The analogue of group IV is a binary compound;

(*e*) the straight line intersects with one of the lines of maximum valence inside the triangle (Fig. 4*e*).

The analogue of group IV is a ternary compound. If we plot all 35 triangular concentration diagrams, we shall verify the correctness of the result obtained above analytically: in five triangles the four-electron straight line will intersect with the "monocationic" line and in another five, with the "dicationic" line.

Quaternary systems. The general formula of a quaternary system will be $A_{(1-x-y-z)}B_xC_yD_z$, where it is assumed, as above, that $A < B < C < D$. The number of combinations of four elements from seven different groups is equal to

$$C_m^n = \frac{7!}{4!\,(7-4)!} = 35.$$

However, depending upon the manner in which the intermediate elements are treated, it now becomes necessary to consider three variants of quaternary systems: the mono-, di- and tri-cationic systems. By the same token, the total number of quaternary systems to be considered thus becomes 105. Among them, we have to select those which are analogues of group IV.

In contrast to ternary systems, where we had two conditions for determining two variables, we now have three variables for the same two conditions. It follows that the quaternary analogue phases of group IV will have a variable composition. We shall illustrate this by considering, for example, the tricationic variant of quaternary systems.

Let us write the condition of normal valence:

$$A(1-x-y-z) + Bx + Cy = (8-D)z$$

and the four-electron condition

$$A(1-x-y-z) + Bx + Cy + Dz = 4.$$

Solving simultaneously, we find

$$x = \frac{8-A-D}{2(B-A)} - \frac{C-A}{B-A}y; \quad y = \frac{8-A-D}{2(C-D)} - \frac{B-A}{C-A}x, \quad z = \frac{1}{2}.$$

Hence we get the formula for the "tricationic" quaternary analogue phase of group IV:

$$A_{(1/2-x-y)}B_xC_yD_{1/2}$$

or

$$A_{\left(1/2 - \frac{8-A-D}{2(B-A)} + \frac{C-A}{B-A}y - y\right)} B_{\left(\frac{8-A-D}{2(B-A)} - \frac{C-A}{B-A}y\right)} C_y D_{1/2}$$

Let us take, for example, the system $1-2-5-6$. Then

$$A = 1; \quad B = 2; \quad C = 5; \quad D = 6; \quad \frac{8-A-D}{2(B-A)} = 1/2; \quad \frac{C-A}{B-A} = 4.$$

FORMATION OF DIAMOND-LIKE SUBSTANCES

The formula of the "tricationic" analogue of elements of group IV in the system 1 – 2 – 5 – 6 assumes the form

$$A^1_{3y} B^2_{(1/2 - 4y)} C^5_y D^6_{1/2}.$$

Since each of the coefficients should be greater than zero, this analogue is possible only if $0 < y < 1/8$. At limiting values of the variable, the quaternary system degenerates into a binary (A^2B^6 at $y = 0$) or ternary ($A^1_3 B^5 C^6_4$ at $y = 1/8$) system. Therefore, the given system may also be represented as a quasi-binary system:

$$m A^2 B^6 \times (1 - m)\, A^1_3 B^5 C^6_4.$$

Table 5 shows all possible quaternary analogues of elements of group IV, which are 37 in number.

It follows from the table that of the 35 combinations, quaternary analogue phases are entirely impossible in four cases, one analogue is obtained in 25 cases, and two analogues in 6 cases.

The same result may be obtained by constructing tetrahedral concentration diagrams (Fig. 5). In this case, each point inside the tetrahedron gives a definite proportion of four elements, the faces represent ternary systems, and the edges, binary ones. Here, the imposition of one of the conditions for the formation of analogues, which reduces the number of the degress of freedom by one, will be represented by the planes of pseudoternary sections. The intersection of two planes will give a straight line which will satisfy both the specified conditions. Depending upon the numbers of the groups which contain the elements comprising a given tetrahedron, the following arrangements of planes are possible:

The four-electron plane

1. is represented by one point at the vertex of the tetrahedron (Fig. 5a); (in this case, the analogue of group IV is the element of group IV itself);

Table 5 Quaternary analogue systems of group IV elements

System	Cation number	General formula	Factorised formula
1–2–3–4	1	—	—
	2	—	—
	3	—	—
1–2–3–5	1	—	—
	2	—	—
	3	—	—
1–2–3–6	1	—	—
	2	—	—
	3	$A^1_x B^2_{(1/2-2x)} C^3_x D^6_{1/2}$	$m A^2 B^6 \times (1-m) A^1 B^3 C^6_2$
1–2–3–7	1	—	—
	2	$A^1_{(4x-3/2)} B^2_{(2-4x)} C^3_{(1/2-x)} D^7_x$	$m A^1 B^7 \times (1-m) A^2_4 B^3 C^7_3$
	3	—	—
1–2–4–5	1	—	—
	2	—	—
	3	$A^1_{(2x-1/2)} B^2_{(1-3x)} C^4_x D^5_{1/2}$	$m A^1 B^4_2 C^5_3 \times (1-m) A^2 B^4 C^5_2$
1–2–4–6	1	—	—
	2	—	—
	3	$A^1_{2x} B^2_{1/2-3x} C^4_x D^6_{1/2}$	$m A^2 B^6 \times (1-m) A^1_2 B^4 C^6_3$
1–2–4–7	1	—	—
	2	$A^1_{3x-1} B^2_{3/2-3x} C^4_{1/2-x} D^7_x$	$m A^1 B^7 \times (1-m) A^2_3 B^4 C^7_2$
	3	—	—
1–2–5–6	1	—	—
	2	—	—
	3	$A^1_{3x} B^2_{1/2-4x} C^5_x D^6_{1/2}$	$m A^2 B^6 \times (1-m) A^1_3 B^5 C^6_4$

Table 5 cont.

System	Cation number	General formula	Factorised formula
1–2–5–7	1	—	—
	2	$A^1_{2x-1} B^2_{1-2x} C^5_{1/2-x} D^7_x$	$mA^1B^7 \times (1-m)\, A^2_2B^5C^7$
	3	—	—
1–2–6–7	1	—	—
	2	$A^1_x B^2_{1/2-x} C^6_{1/2-x} D^7_x$	$mA^1B^7 \times (1-m)\, A^2B^6$
	3	—	—
1–3–4–5	1	—	—
	2	—	—
	3	$A^1_{1/2x} B^3_{1/2-3/2x} C^4_x D^5_{1/2}$	$mA^8B^5 \times (1-m)\, A^1B^4_2C^5_3$
1–3–4–6	1	—	—
	2	$A^1_{x-1/4} B^3_{3/4-x} C^4_{1/2-x} D^6_x$	$mA^1B^3C^6_2 \times (1-m)\, A^3_2B^4C^6$
	3	$A^1_{1/4+1/2x} B^3_{1/4-3/2x} C^4_x D^6_{1/2}$	$mA^1B^3C^6_2 \times (1-m)\, A^3_2B^4C^6_3$
1–3–4–7	1	—	—
	2	$A^1_{3/2x-1/4} B^3_{3/4-3/2x} C^4_{1/2-x} D^7_x$	$mA^1B^7 \times (1-m)\, A^3_3B^4_2C^7$
	3	—	—
1–3–5–6	1	—	—
	2	$A^1_{1/2x} B^3_{1/2-1/2x} C^5_{1/2-x} D^6_x$	$mA^3B^5 \times (1-m)\, A^1B^8C^6_2$
	3	$A^1_{1/4+x} B^3_{1/4-2x} C^5_x D^6_{1/2}$	$mA^1B^3C^6_2 \times (1-m)\, A^1_3B^5C^6_4$
1–3–5–7	1	—	—
	2	$A^1_x B^3_{1/2-x} C^5_{1/2-x} D^7_x$	$mA^1B^7 \times (1-m)\, A^8B^5$
	3	—	—
1–3–6–7	1	—	—
	2	$A^1_{1/4+1/2x} B^3_{1/4-1/2x} C^5_{1/2-x} D^7_x$	$mA^1B^7 \times (1-m)\, A^1B^3C^6_2$
	3	—	—

Table 5 cont.

System	Cation number	General formula	Factorised formula
1–4–5–6	1	—	—
	2	$A^1_{1/6+1/3x}B^4_{1/3-1/3x}C^5_{1/2-x}D^6_x$	$mA^1B^4_2C^5_3 \times (1-m)\,A^1_2B^4C^6_3$
	3	$A^1_{1/3+1/3x}B^4_{1/6-1/3x}C^5_xD^6_{1/2}$	$mA^1_2B^4C^6_3 \times (1-m)\,A^1_3B^5C^6_4$
1–4–5–7	1	—	—
	2	$A^1_{1/6+2/3x}B^4_{1/3-2/3x}C^5_{1/2-x}D^7_x$	$mA^1B^7 \times (1-m)A^1B^4_2C^5_3$
	3	—	—
1–4–6–7	1	—	—
	2	$A^1_{1/3+1/3x}B^4_{1/6-1/3x}C^6_{1/2-x}D^7_x$	$mA^1B^7 \times (1-m)\,A^1_2B^4C^6_3$
	3	—	—
1–5–6–7	1	—	—
	2	$A^1_{3/8+1/4x}B^5_{1/8+1/4x}C^6_{1/2-x}D^7_x$	$mA^1B^7 \times (1-m)\,A^1_3B^5C^6_4$
	3	—	—
2–3–4–5	1	—	—
	2	—	—
	3	$A^2_xB^3_{1/2-2x}C^4_xD^5_{1/2}$	$mA^3B^5 \times (1-m)\,A^2B^4C^5_2$
2–3–4–6	1	—	—
	2	$A^2_{2x-1/2}B^3_{1-2x}C^4_{1/2-x}D^6_x$	$mA^2B^6 \times (1-m)\,A^3_2B^4C^6$
	3	—	—
2–3–4–7	1	$A^2_{1/2}B^3_{3x-1}C^4_{3/2-4x}D^7_x$	$mA^2_3B^4C^7_2 \times (1-m)\,A^2_4B^3C^7_3$
	2	$A^2_{3x-1/2}B^3_{1-3x}C^4_{1/2-x}D^7_x$	$mA^2_3B^4C^7_2 \times (1-m)\,A^3_3B^4_2C^7$
	3	—	—
2–3–5–6	1	—	—
	2	$A^2_xB^3_{1/2-x}C^5_{1/2-x}D^6_x$	$mA^2B^6 \times (1-m)\,A^3B^5$
	3	—	—

Table 5 cont.

System	Cation number	General formula	Factorised formula
2–3–5–7	1	$A^2_{1/2}B^3_{x-1/2}C^5_{3/4-2x}D^7_x$	$mA^2_2B^5C^7 \times (1-m)\ A^2_4B^3C^7_3$
	2	$A^2_{2x}B^3_{1/2-2x}C^5_{1/2-x}D^7_x$	$mA^3B^5 \times (1-m)\ A^2_2B^5C^7$
	3	—	—
2–3–6–7	1	$A^2_{1/2}B^3_{1/2x}C^6_{1/2-4/3x}D^7_x$	$mA^2B^6 \times (1-m)\ A^2_4B^3C^7_3$
	2	—	—
	3	—	—
2–4–5–6	1	—	—
	2	$A^2_{1/2+1/2x}B^4_{1/4-1/2x}C^5_{1/2-x}D^6_x$	$mA^2B^6 \times (1-m)\ A^2B^4C^5_2$
	3	—	—
2–4–5–7	1	$A^2_{1/2}B^4_{2x-1/2}C^5_{1-3x}D^7_x$	$mA^2_2B^5C^7 \times (1-m)\ A^2_3B^4C^7_2$
	2	$A^2_{1/4+x}B^4_{1/4-x}C^5_{1/2-x}D^7_x$	$mA^2B^4C^5_2 \times (1-m)\ A^2_2B^5C^7$
	3	—	—
2–4–6–7	1	$A^2_{1/2}B^5_{1/2x}C^6_{1/2-3/2x}D^7_x$	$mA^2B^6 \times (1-m)\ A^2_3B^4C^7_2$
	2	—	—
	3	—	—
2–5–6–7	1	$A^2_{1/2}B^5_xC^6_{1/2-2x}D^7_x$	$mA^2B^6 \times (1-m)\ A^2_2B^5C^7$
	2	—	—
	3	—	—
3–4–5–6	1	$A^3_{1/2}B^4_xC^5_{1/2-2x}D^6_x$	$mA^3B^5 \times (1-m)\ A^3_2B^4C^6$
	2	—	—
	3	—	—
3–4–5–7	1	$A^3_{1/2}B^4_{2x}C^5_{1/2-3x}D^7_x$	$mA^3B^5 \times (1-m)\ A^3_3B^4_2C^7$
	2	—	—
	3	—	—

Table 5 cont.

System	Cation number	General formula	Factorised formula
3–4–6–7	1	$A^3_{1/2}B^4_{1/4+1/2x}C^6_{1/4-3/2x}D^7_x$	$mA^3_2B^4C^6x(1-m)\,A^3_3B^4_2C^7$
	2	—	—
	3	—	—
3–5–6–7	1	—	—
	2	—	—
	3	—	—
4–5–6–7	1	—	—
	2	—	—
	3	—	—

2. exists and has a common point with one of the planes of completed valence at the edge of the tetrahedron (Fig. 5b), in which case a binary phase will be the only analogue of group IV;

3. intersects with one of the planes of maximum valence, the line of intersection ending: (a) at the edges of the tetrahedron (Fig. 5c): the quaternary analogue system is made up of two binary systems; (b) at the faces of the tetrahedron (Fig. 5d): the quaternary analogue system is made up of two ternary systems; (c) at the edge and at the face (Fig. 5e): the quaternary analogue system is made up of a double and a triple system;

4. intersects with two planes of maximum valence. Here two analogue systems are obtained which have one common component. Depending upon the positions of the ends of the lines of intersection, two cases are possible: (a) both systems are made up of two ternary

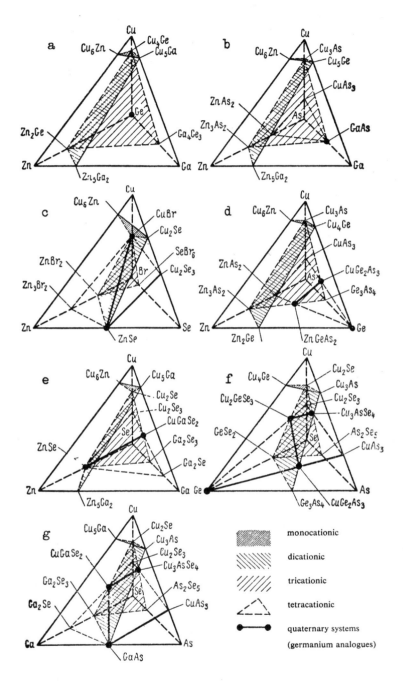

Fig. 5 *Formation of quaternary tetrahedral analogue phases of germanium*

systems (Fig. 5f); (b) one system is made up of two ternary ones and the second consists of one binary and one ternary system (Fig. 5g).

Quinary systems. The general formula of quinary systems is

$$A_{(1-x-y-z-u)}B_xC_yD_zE_u,$$

where it is assumed that A < B < C < D < E.

The number of combinations of five elements from seven different groups is equal to

$$C_m^n = \frac{7!}{5!\,(7-5)!} = 21,$$

and the number of variables is four, so that 84 variants of systems must be taken into consideration.

We now have the same two conditions to find four unknowns. It follows that the quinary analogue phases of group IV will have a variable composition with two degrees of freedom.

To find the quinary analogues of elements of group IV, we shall write the condition of complete valence for each of the variants:

(*a*) the "monocationic" variant

$$A(1-x-y-z-u) = (8-B)x + (8-C)y + (8-D)z + (8-E)u,$$

(*b*) the "dicationic" variant

$$A(1-x-y-z-u) + Bx = (8-C)y + (8-D)z + (8-E)u,$$

(*c*) the "tricationic" variant

$$A(1-x-y-z-u) + Bx + Cy = (8-D)z + (8-E)u,$$

(*d*) the "tetracationic" variant

$$A(1-x-y-z-u) + Bx + Cy + Dz = (8-E)u.$$

Solving each together with the condition that the number of valence electrons per atom be four

$$A(1-x-y-z-u) + Bx + Cy + Dz + Eu = 4,$$

we obtain the formulae for quinary analogue phases of elements of group IV, a complete list of which, including 42 systems, is given in Table 6.

The region of existence of each system is defined by the positive values of the coefficients. For example, in the case of the "dicationic" variant of the system 1 - 3 - 4 - 5 - 6, whose formula is

$$A^1_{-1/4+1/2x+y} B^3_{3/4-1/2x-y} C^4_{1/2-x-y} D^5_x E^6_y,$$

the region of existence is bounded by the three intersecting lines

$$-1/4 + 1/2 x + y = 0$$
$$1/2 - x - y = 0$$
$$x = 0,$$

since the two remaining conditions

$$3/4 - 1/2 x - y = 0$$
$$y = 0$$

define a region within which that bounded by the three lines wholly lies. These two conditions, therefore, are satisfied identically (Fig. 6a). The boundaries correspond to the quaternary systems

$$mA^3B^5 \cdot (1-m) A^3_2 B^4 C^6$$
$$mA^3B^5 \cdot (1-m) A^1 B^3 C^6_2$$
$$mA^1B^3C^6_2 \cdot (1-m) A^3_2 B^4 C^6,$$

Table 6 Quinary analogue systems of group IV elements

System	Cation number	Formula	No. of co-existing phases Quaternary	No. of co-existing phases Ternary and binary
1–2–3–4–5	1	—	—	—
	2	—	—	—
	3	—	—	—
	4	$A^1_{-1/2 + x + 2y} B^2_{1 - 2x - 3y} C^3_x D^4_y E^5_{1/2}$	3	3
1–2–3–4–6	1	—	—	—
	2	—	—	—
	3	$A^1_{-1 + x + 2y} B^2_{3/2 - 2x - 2y} C^3_x D^4_{1/2 - y} E^6_y$	3	3
	4	$A^1_{x + 2y} B^2_{1/2 - 2x - 3y} C^3_x D^4_y E^5_{1/2}$	3	3
1–2–3–4–7	1	—	—	—
	2	$A^1_{-3/2 + x + 4y} B^2_{2 - x - 4y} C^3_{1/2 - x - y} D^4_x E^7_y$	3	3
	3	$A^1_{-1 + x + 3y} B^2_{3/2 - 2x - 3y} C^3_x D^4_{1/2 - y} E^7_y$	3	3
	4	—	—	—
1–2–3–5–6	1	—	—	—
	2	—	—	—
	3	$A^1_{-1/2 + x + y} B^2_{1 - 2x - y} C^3_x D^5_{1/2 - y} E^6_y$	3	3
	4	$A^1_{x + 3y} B^2_{1/2 - 2x - 4y} C^3_x D^5_y E^6_{1/2}$	3	3
1–2–3–5–7	1	—	—	—
	2	$A^1_{-3/2 + 2x + 4y} B^2_{2 - 2x - 4y} C^3_{1/2 - x - y} D^5_x E^7_y$	3	3
	3	$A^1_{-1/2 + x + 2y} B^2_{1 - 2x - 2y} C^3_x D^5_{1/2 - y} E^7_y$	3	3
	4	—	—	—
1–2–3–6–7	1	—	—	—
	2	$A^1_{-3/2 + 3x + 4y} B^2_{2 - 3x - 4y} C^3_{1/2 - x - y} D^6_x E^7_y$	3	3
	3	$A^1_{x - y} B^2_{1/2 - 2x - y} C^3_x D^6_{1/2 - y} E^7_y$	3	3
	4	—	—	—

Table 6 cont.

System	Cation number	Formula	No. of co-existing phases	
			Quaternary	Ternary and binary
1–2–4–5–6	1	—	—	—
	2	—	—	—
	3	$A^1_{-1/2 + 2x + y} B^2_{1 - 3x - y} C^4_x D^5_{1/2 - y} E^6_y$	4	4
	4	$A^1_{2x + 3y} B^2_{1/2 - 3x - 4y} C^4_x D^5_y E^6_{1/2}$	3	3
1–2–4–5–7	1	—	—	—
	2	$A^1_{-1 + x + 3y} B^2_{3/2 - x - 3y} C^4_{1/2 - x - y} D^5_y E^7_y$	3	3
	3	$A^1_{-1/2 + 2x + 2y} B^2_{1 - 3x - 2y} C^4_x D^5_{1/2 - y} E^7_y$	4	4
	4	—	—	—
1–2–4–6–7	1	—	—	—
	2	$A^1_{-1 + 2x + 3y} B^2_{3/2 - 2x - 3y} C^4_{1/2 - x - y} D^6_x E^7_y$	3	3
	3	$A^1_{2x + y} B^2_{1/2 - 3x - y} C^4_x D^6_{1/2 - x} E^7_y$	3	3
	4	—	—	—
1–2–5–6–7	1	—	—	—
	2	$A^1_{-1/2 + x + 2y} B^2_{1 - x - 2y} C^5_{1/2 - x - y} D^6_x E^7_y$	3	3
	3	$A^1_{3x + y} B^2_{1/2 - 4x - y} C^5_x D^6_{1/2 - y} E^7_y$	3	3
	4	—	—	—
1–3–4–5–6	1	—	—	—
	2	$A^1_{-1/4 + 1/2 x + y} B^3_{3/4 - 1/2 x - y} C^4_{1/2 - x - y} D^5_x E^6_y$	3	3
	3	$A^1_{1/2 x + 1/2 y} B^3_{3/2 - 3/2 x - 1/2 y} C^4_x D^5_{1/2 - y} E^6_y$	4	4
	4	$A^1_{1/4 + 1/2 x + y} B^3_{1/4 - 3/2 x - 2y} C^4_x D^5_y E^6_{1/2}$	3	3
1–3–4–5–7	1	—	—	—
	2	$A^1_{-1/4 + 1/2 x + 3/2 y} B^3_{3/4 - 1/2 x - 3/2 y} C^4_{1/2 - x - y} D^5_x E^7_y$	3	3
	3	$A^1_{1/2 x + y} B^3_{1/2 - 3/2 x - y} C^4_x D^5_{1/2 - y} E^7_y$	3	3
	4	—	—	—

Table 6 cont.

System	Cation number	Formula	No. of co-existing phases	
			Quaternary	Ternary and binary
1–3–4–6–7	1	—	—	—
	2	$A^1_{-1/4 + x + 3/2 y} B^3_{3/4 - x - 3/2 y} C^4_{1/2 - x - y} D^6_x E^7_y$	4	4
	3	$A^1_{1/4 + 1/2 x + 1/2 y} B^3_{1/4 - 3/2 x - 1/2 y} C^4_x D^6_{1/2 - y} E^7_y$	3	3
	4	—	—	—
1–3–5–6–7	1	—	—	—
	2	$A^1_{1/2 x + y} B^3_{1/2 - 1/2 x - y} C^5_{1/2 - x - y} D^6_x E^7_y$	3	3
	3	$A^1_{1/4 + x + 1/2 y} B^3_{1/4 - 2x - 1/2 y} C^5_x D^6_{1/2 - y} E^7_y$	3	3
	4	—	—	—
1–4–5–6–7	1	—	—	—
	2	$A^1_{1/6 + 1/3 x + 2/3 y} B^4_{1/3 - 1/3 x - 2/3 y} C^5_{1/2 - x - y} D^6_x E^7_y$	3	3
	3	$A^1_{1/3 + 1/3 x + 1/3 y} B^4_{1/6 - 4/3 x - 1/3 y} C^5_x D^6_{1/2 - y} E^7_y$	3	3
	4	—	—	—
2–3–4–5–6	1	—	—	—
	2	$A^2_{-1/2 + x + 2y} B^3_{1 - x - 2y} C^4_{1/2 - x - y} D^5_x E^6_y$	3	3
	3	$A^2_{x + y} B^3_{1/2 - 2x - y} C^4_x D^5_{1/2 - y} E^6_y$	3	3
	4	—	—	—
2–3–4–5–7	1	$A^2_{1/2} B^3_{-1 + x + 3y} C^4_{3/2 - 2x - 4y} D^5_x E^7_y$	3	3
	2	$A^2_{-1/2 + x + 3y} B^3_{1 - x + 3y} C^4_{1/2 - x - y} D^5_x E^7_y$	4	4
	3	$A^2_{x + 2y} B^3_{1/2 - 2x - 2y} C^4_x D^5_{1/2 - y} E^7_y$	3	3
	4	—	—	—
2–3–4–6–7	1	$A^2_{1/2} B^3_{-1 + 2x + 3y} C^4_{3/2 - 3x - 4y} D^6_x E^7_y$	3	3
	2	$A^2_{-1/2 + 2x + 3y} B^3_{1 - 2x - 3y} C^4_{1/2 - x - y} D^6_x E^7_y$	3	3
	3	—	—	—
	4	—	—	—

Table 6 cont.

System	Cation number	Formula	No. of co-existing phases	
			Quaternary	Ternary and binary
2–3–5–6–7	1	$A^2_{1/2}B^3_{-1/4+1/2x+y}C^5_{3/4-3/2x-2y}D^6_xE^7_y$	3	3
	2	$A^2_{x+2y}B^3_{1/2-x-2y}C^5_{1/2-x-y}D^6_xE^7_y$	3	3
	3	—	—	—
	4	—	—	—
2–4–5–6–7	1	$A^2_{1/2}B^4_{-1/2+x+2y}C^5_{1-2x-3y}D^6_xE^7_y$	3	3
	2	$A^2_{1/4+1/2x+y}B^4_{1/4-1/2x-y}C^5_{1/2-x-y}D^6_xE^7_y$	3	3
	3	—	—	—
	4	—	—	—
3–4–5–6–7	1	$A^3_{1/2}B^4_{x+2y}C^5_{1/2-2x-3y}D^6_xE^7_y$	3	3
	2	—	—	—
	3	—	—	—
	4	—	—	—

and the intersections of the boundaries correspond to binary (A^3B^5 at $x = 1/2$; $y = 0$) or ternary ($A^1B^3C^6_2$ at $x = 0$, $y = 1/2$ and $A^3_2B^4C^6$ at $x = 0$, $y = 1/4$) systems.

Therefore, this system (the "dicationic" variant of the system 1 – 3 – 4 – 5 – 6) can also be represented in the form of a quasiternary system:

$$mA^8B^5 \cdot nA^1B^3C_2^6 \cdot (1-m-n)\, A_2^3B^4C^6.$$

For a "tricationic" variant of the same system, which has the formula

$$A^1_{1/2 x + 1/2 y}\, B^3_{1/2 - 3/2 x - 1/2 y}\, C^4_x D^5_{1/2 - y}\, E^6_y,$$

the region of existence is bounded by the four straight lines (Fig. 6b)

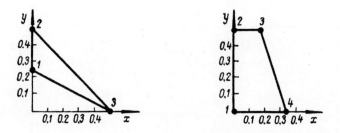

Fig. 6 *Various types of quinary tetrahedral systems*

a $\quad A^1_{\frac{1}{4}+\frac{1}{2}x+y}\, B^3_{\frac{3}{4}-\frac{1}{2}x-y}\, C^4_{\frac{1}{2}-x-y}\, D^5_x E^6_y,$

b $\quad A^1_{\frac{1}{2}x+\frac{1}{2}y}\, B^3_{\frac{1}{2}-\frac{3}{4}x+\frac{1}{2}y}\, C^4_x D^5_{\frac{1}{2}-y}\, E^6_y$

$$1/2 - 3/2 x - 1/2 y = 0$$
$$x = 0$$
$$1/2 - y = 0$$
$$y = 0.$$

Therefore, at the boundaries, the quinary system degenerates into one of the following four quaternary ones:

$$mA_2^1B^4C_3^6 \cdot (1-m) A^1B_2^4C_3^5$$
$$mA^3B^5 \cdot (1-m) A^1B^8C_2^6$$
$$mA^1B^8C_2^6 \cdot (1-m) A_2^1B^4C_3^6$$
$$mA^3B^5 \cdot (1-m) A^1B_2^4C_3^5,$$

made up of binary (A^3B^5 at $x = 0$, $y = 0$) or ternary ($A^1B^3C_2^6$ at $x = 0$, $y = 1/2$; $A^1B_2^4C_3^5$ at $x = 1/3$, $y = 0$ and $A_2^1B^4C_3^6$ at $x = 1/6$, $y = 1/2$) systems.

In this case, the quinary system can be represented by binary and ternary systems in various ways:

1. By two of the opposite angles of the region of existence (for example, $m3A^3B^5 (1-m) A_2^1B^4C_3^6$ on the corresponding curve),

a $\quad A_{1+x-2y-z}^2 \quad B_{\frac{1}{2}-2x+2y+z}^3 \quad C_x^4D_y^5E_z^6F_{\frac{1}{2}-y-z}^7$

b $\quad A_{\frac{1}{2}+\frac{1}{2}x-y-\frac{1}{2}z}^1 \quad B_{\frac{1}{2}-\frac{3}{2}x+y+\frac{1}{2}x}^3 \quad C_x^4D_y^5E_z^6F_{\frac{1}{2}-y-z}^7$

c $\quad A_{\frac{1}{2}+2x-2y-z}^1 \quad B_{-3x+2y+z}^2 \quad C_x^4D_y^5E_z^6F_{\frac{1}{2}-y-z}^7$

d $\quad A_{\frac{1}{2}+x-2y-z}^1 \quad B_{-2x+2y+z}^2 \quad C_x^3D_y^5E_z^6F_{\frac{1}{2}-y-z}^7$

2. By three of the adjacent angles (for example, $m3A^3B^5nA^1B_2^4C_3^5$ $(1-m-n) A_2^1B^4C_3^6$ on the corresponding part of the region of existence),

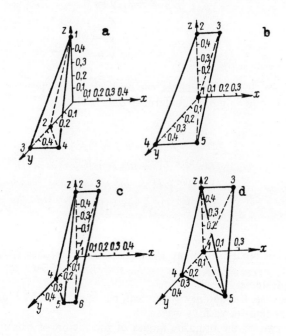

Fig. 7 *Various types of sextuple tetrahedral systems*

3. in general, by all four:

$$n\,[m6A^3B^5 \cdot (1-m)\,3A^1B^3C_2^6] \times$$
$$\times (1-n)\,[m2A^1B_2^4C_3^5 \cdot (1-m)\,2A_2^1B^4C_3^6].$$

Therefore, Table 6 and the following tables of this section give only the basic formulae of complex systems and the quantities of simpler phases into which they degenerate at the boundaries of the region of existence.

Sextuple systems. All the above considerations may be applied to the identification of six-component systems of analogues of elements

FORMATION OF DIAMOND-LIKE SUBSTANCES 37

in group IV.

To this end, from the general formula of a sextuple system

$$A_{1-x-y-z-u-v}B_xC_yD_zE_uF_v,$$

using the two conditions of formation, we eliminate two unknowns, and thus obtain formulae of analogues with three degrees of freedom.

The region of existence of each of them, as above, is defined by the positive values of the coefficients. Depending upon the number of simpler phases into which the sextuple system degenerates at the boundaries of the region of existence, four cases are possible, examples of which are illustrated graphically in Fig. 7.

In the first case (Fig. 7a), the three-dimensional shape representing the region of existence of the sextuple system is bounded by four planes representing quinary systems, six edges representing quaternary systems, and four vertices representing ternary or binary compounds.

In the second case (Fig. 7b) we have, respectively, five quinary, eight quaternary and five ternary or binary systems; in the third case (Fig. 7c), six quinary, ten quaternary and six ternary or binary systems; in the fourth case (Fig. 7d), six quinary, nine quaternary and five ternary or binary systems.

The corresponding data for each sextuple analogue system, the total number of these systems being 21, are given in Table 7.

Seven-component systems. By carrying out similar calculations for seven-component systems, we shall obtain formulae for seven-component analogues whose composition has four degrees of freedom. Accordingly, the region of their existence will also constitute a volume in four-dimensional space. In this case also, depending upon the number of the simpler phases into which the seven-component system degenerates at the boundaries of the region of existence, two cases are possible. In the first, the four-dimensional region is bounded by five three-dimensional ones representing sextuple systems, ten two-dimensional ones representing quinary systems, ten uni-dimensional ones and five zero-dimensional ones which are ternary or binary phases. In the second case, this region is bounded, respectively, by seven three-dimensional, twenty two-dimensional, nineteen uni-dimensional and eight zero-dimensional phases.

The total number of seven-component analogue systems is four. The data for each of them are shown in Table 8.

By these methods, we have determined the number of possible

Table 7 Sextuple analogue systems of group IV elements

System	Cation number	Formula	No. of co-existing phases		
			Quinary	Quaternary	Ternary and binary
1–2–3–4–5–6	1	—	—	—	—
	2	—	—	—	—
	3	$A^1_{x-2y-z}B^2_{1/2-2x+2y+z}C^3_xD^4_yE^5_zF^6_{1/2-y-z}$	4	6	4
	4	$A^1_{x+2y-z}B^2_{1/2-2x-3y+z}C^3_xD^4_yE^5_zF^6_{1/2-z}$	6	10	6
	5	$A^1_{x+2y+3z}B^2_{1/2-2x-3y-4z}C^3_xD^4_yE^5_zF^6_{1/2}$	4	6	4
1–2–3–4–5–7	1	—	—	—	—
	2	$A^1_{1/2-4x-3y-2z}B^2_{4x+3y+2z}C^3_xD^4_yE^5_zF^7_{1/2-x-y-z}$	4	6	4
	3	$A^1_{1/2+x-3y-3z}B^2_{-2x+3y+2z}C^3_xD^4_yE^5_zF^7_{1/2-y-z}$	5	8	5
	4	$A^1_{1/2+x+2y-2z}B^2_{-2x-3y+2z}C^3_xD^4_yE^5_zF^7_{1/2-z}$	5	8	5
	5	—	—	—	—
1–2–3–4–6–7	1	—	—	—	—
	2	$A^1_{1/2-4x-3y-z}B^2_{4x+3y+z}C^3_xD^4_yE^6_zF^7_{1/2-x-y-z}$	4	6	4
	3	$A^1_{1/2+x-3y-z}B^2_{-2x+3y+z}C^3_xD^4_yE^6_zF^7_{1/2-y-z}$	6	10	6
	4	$A^1_{1/2+x+2y-z}B^2_{-2x-3y+z}C^3_xD^4_yE^6_zF^7_{1/2-z}$	4	6	4
	5	—	—	—	—
1–2–3–5–6–7	1	—	—	—	—
	2	$A^1_{1/2-4x-2y-z}B^2_{4x+2y+z}C^3_xD^5_yE^6_zF^7_{1/2-x-y-z}$	4	6	4
	3	$A^1_{1/2+x-2y-z}B^2_{-2x+2y+z}C^3_xD^5_yE^6_zF^7_{1/2-y-z}$	6	9	5
	4	$A^1_{1/2+x+3y-z}B^2_{-2x-4y+z}C^3_xD^5_yE^6_zF^7_{1/2-z}$	4	6	4
	5	—	—	—	—
1–2–4–5–6–7	1	—	—	—	—
	2	$A^1_{1/2-3x-2y-z}B^2_{3x+2y+z}C^4_xD^5_yE^6_zF^7_{1/2-x-y-z}$	4	6	4
	3	$A^1_{1/2+2x-2y-z}B^2_{-3x+2y+z}C^4_xD^5_yE^6_zF^7_{1/2-y-z}$	6	10	6
	4	$A^1_{1/2+2x+3y-z}B^2_{-3x-4y+z}C^4_xD^5_yE^6_zF^7_{1/2-z}$	4	6	4
	5	—	—	—	—

Table 7 cont.

System	Cation number	Formula	No. of co-existing phases		
			Quinary	Quaternary	Ternary and binary
1–3–4–5–6–7	1	—	—	—	—
	2	$A^1_{1/2 - 3/2x - y - 1/2z} B^3_{3/2x + y + 1/2z} C^4_x D^5_y E^6_z F^7_{1/2 - x - y - z}$	5	8	5
	3	$A^1_{1/2 + 1/2x - y - 1/2z} B^3_{3/2x + y + 1/2z} C^4_x D^5_y E^6_z F^7_{1/2 - y - z}$	5	8	5
	4	$A^1_{1/2 + 1/2x + y - 1/2z} B^3_{-3/2x - 2y + 1/2z} C^4_x D^5_y E^6_z F^7_{1/2 - z}$	4	6	4
	5	—	—	—	—
2–3–4–5–6–7	1	$A^2_{1/2} B^3_{1/8 - 3/4x - 1/2y - 1/4z} C^4_x D^5_y E^6_z F^7_{3/8 - 1/4x - 1/2y - 3/4z}$	4	6	4
	2	$A^2_{1 - 3x - 2y - z} B^3_{-1/2 + 3x + 2y + z} C^4_x D^5_y E^6_z F^7_{1/2 - x - y - z}$	6	10	6
	3	$A^2_{1 + x - 2y - z} B^3_{-1/2 - 2x + 2y + z} C^4_x D^5_y E^6_z F^7_{1/2 - y - z}$	4	6	4
	4	—	—	—	—
	5	—	—	—	—

Table 8 Seven-component analogue systems of group IV elements

System	Cation number	Formula	No. of co-existing phases			
			Sextuple	Quinary	Quaternary	Ternary and binary
1–2–3–4–5–6–7	1	—	—	—	—	—
	2	$A^1_{1/2 - 4x - 3y - 2z - u} B^2_{4x + 3y + 2z + u} C^3_x D^4_y E^5_z F^6_u G^7_{1/2 - x - y - z - u}$	5	10	10	5
	3	$A^1_{1/2 + x - 3y - 2z - u} B^2_{-2x + 3y + 2z + u} C^3_x D^4_y E^5_z F^6_u G^7_{1/2 - y - z - u}$	7	20	19	8
	4	$A^1_{1/2 + x + 2y - 2z - u} B^2_{-2x - 3y + 2z + u} C^3_x D^4_y E^5_z F^6_u G^7_{1/2 - z - u}$	7	20	19	8
	5	$A^1_{1/2 + x + 2y + 3z - u} B^2_{-2x - 3y - 4z + u} C^3_x D^4_y E^5_z F^6_u G^7_{1/2 - u}$	5	10	10	5
	6	—	—	—	—	—

types of tetrahedral phases. However, this method takes into consideration only their common feature, which is that they all have the same average number of valence electrons per atom; it also takes into account the valence type.

However, if we proceed from the general formula to specific elements, we also have to take into consideration their positions in the different periods of the periodic system, these positions being related to the number of filled electron shells; appreciable corrections are thereby introduced into the prospects of obtaining any given tetrahedral phase.

It is known, for example, that in group IV only four out of five elements, or 80%, crystallize in the structure of diamond. Of the binary compounds closest to it, 15 out of 25 or 60% of type A^3B^5 and 60% of type A^2B^6 give this structure, whilst in compounds of type A^1B^7 the number of tetrahedral phases which actually exist drops to 16%. Likewise, only a few of the possibilities in more complex systems may actually exist. However, as will be shown below, the impossibility of realizing simpler phases does not exclude their participation in more complex ones. For instance, in the system $m\text{InSb} \cdot (1-m)\ \text{CdSnSb}_2$ there is an extensive region of homogeneous solutions, while CdSnSb_2 does not exist as a separate substance[20].

The best agreement with the predictions based on the above considerations is shown by substances comprised of elements located in the fourth period, and, to a greater or lesser extent, in the third and fifth period. As a rule, substances in the "diagonal" isoelectronic series, including the fourth period, also conform to this regularity (see Chapter 2). Systems made up of elements of the sixth or the sixth and fifth periods usually give deviations from the rules due to the stability of the s^2 subshell of these elements.

To date, complex tetrahedral phases up to and including sextuple ones have been experimentally shown to exist. The prospects of obtaining new semiconducting materials have thereby been expanded immeasurably.

In analysing the general conclusions reached on the basis of the above material, it is necessary to keep in mind that the discussion completely omitted phases in which the elements of the same combination played the part of "cations" and "anions" simultaneously.

What is the probability of this happening in tetrahedral phases? Obviously, for substances with a pronounced covalent bond, in which the electronic lattice determines the structure and properties of the substances to a greater extent than the composition of the lattice points[5], this probability is greater than for ionic compounds.

Experience has shown that certain elements of groups IV and V

FORMATION OF DIAMOND-LIKE SUBSTANCES

of the periodic system may indeed play the parts of "cations" and "anions" in complex tetrahedral phases simultaneously, i.e., by filling up their valence shell by accepting up to 8 electrons or by giving them up completely.

Thus it was found that in ternary tetrahedral phases containing germanium, such as $CuGe_2P_3$, $ZnGeAs_2$ and Cu_2GeSe_3, germanium may dissolve in large amounts. X-ray studies have shown that germanium occupies sites in the lattice which correspond not only to those of copper and zinc but also to the sites of non-metals[21].

In the system $m4InSb \cdot (1-m)$ Ag_3SbTe_4 in the vicinity of InSb, there are homogeneous alloys in which the sites of the "anions" are apparently occupied by antimony and tellurium, and those of the "cations", by silver, indium and antimony[22, 23]. Thus, antimony and indium are found to occupy crystallographically equivalent positions, something that is not observed in binary tetrahedral phases. This fact suggests that variations in the composition of the lattice points can apparently be more extensive in complex tetrahedral phases than in simple ones.

Owing to this departure from the conditions initially adopted, i.e., to the fact that the elements of groups IV and V may play the part of "cations" and "anions" simultaneously, the analytical and graphic solution becomes more complicated for certain cases. For example, in the concentration triangle "2 – 4 – 5", the ternary compound designated by a point (for the isoelectronic series of germanium $ZnGeAs_2$)—and for this reason having a small region of homogeneity—should be joined by a line with the angle of the triangle where germanium is located. This signified a possible formation of homogeneous alloys along this section, i.e., the conversion of the chemical compound $ZnGeAs_2$ into a phase of variable composition with respect to one of the components.

Similarly, in quaternary systems involving elements of group IV, the regions of homogeneity may not be bounded by straight lines but extend to the four-electron plane; this may be treated as a variable valence of the element of group IV.

In the following section we shall examine some defect tetrahedral phases, and in Part D of Chapter 2, their numerous solid solutions with non-defect substances.

Therefore, at this point, while discussing ternary and quaternary systems— and running a little ahead—we shall note that in considering the regions of homogeneity, we should keep in mind the possible formation of homogeneous regions with defect tetrahedral structures in these phases.

This type of approach leads to more accurate predictions than the empirical analysis of a quaternary system given in reference 24, where only a very small region of homogeneity was determined experimentally in the system Zn – Ga – Si – P. This region actually appears to be considerably wider, as in other similar quaternary systems.

3. Defect and Excess Tetrahedral Phases

In the preceding section we presented a detailed discussion of complex tetrahedral phases possessing an average of four electrons per atom. However, experiments show that substances with a number of electrons different from four also possess semiconducting properties. Thus, even certain elements of group III (boron), and modifications of elements of groups V (phosphorus, arsenic, antimony), VI (sulphur, selenium, tellurium) and even VII (iodine), manifest such properties and for this reason can in a certain sense be classed with the elements of group IV.

Among binary compounds one finds a whole group of semiconducting substances of the type $A_2^3B_3^6$ with a tetrahedral structural arrangement of the atoms, the only difference being that one-third of the lattice points occupied by the atoms which are "cations" remain vacant, and are therefore termed defect lattice points. We can also cite examples of excess tetrahedral phases (Cu_2Se, etc.) where the atoms which are "cations" occupy all the tetrahedral vacancies of the cubic packing of the "anions", as well as phases of the type $A_3^1B^5$ (Li_3Bi, etc.) where the octahedral vacancies are also occupied.

All this makes it possible to treat the group of diamond-like semiconductors together with the crystallochemically related defect and excess tetrahedral phases having an average number of valence electrons per atom different from four. Hence the question arises as to the limits, between which the average number of electrons per atom can change, where one can still expect the formation of defect or excess tetrahedral phases with semiconducting properties. The concomitant question is what are the possible types of defect or excess tetrahedral complex substances with a (basically) tetrahedral structural arrangement of atoms?

To answer the first question, we shall set up the formulae of all the possible binary compounds of normal valence and write them out, along with the elements of the seven groups participating in the formation of tetrahedral phases, in the order of increasing average number of electrons per atom (Table 9). Some of the binary compounds included in the table do not exist individually but, as was noted above, they may be present as components of more complex phases and therefore should not be excluded from consideration.

It is evident from Table 9 that the binary analogues of group II elements are substances of the type $A_3^1B^5$ (Cu_3As). Among the binary compounds of normal valence there are no analogues of group III elements. Germanium has three types of analogue: A^3B^5, A^2B^6 and A^1B^7, as has already been shown experimentally. The binary analogues of selenium are substances of the type of $GaBr_3$.

The latter analogue has not been studied at all, any more than has

Table 9 Elements and binary compounds of normal valence

General Formula	Element or compound	Number of el/at	General Formula	Element or compound	Number of el/at
A^1	Cu	1	A^2B^6	ZnSe	4.00
$A_6^1B^2$	Cu_6Zn	1.14	A^3B^5	GaAs	4.00
$A_5^1B^3$	Cu_5Ga	1.33	$A_3^4B_4^5$	Ge_3As_4	4.57
$A_4^1B^4$	Cu_4Ge	1.60	$A_2^3B_3^6$	Ga_2Se_3	4.80
A^2	Zn	2.00	A^5	As	5.00
$A_3^1B^5$	Cu_3As	2.00	$A^2B_2^7$	$ZnBr_2$	5.33
$A_5^2B_2^3$	Zn_5Ga_2	2.29	$A^4B_2^6$	$GeSe_2$	5.33
$A_2^1B^6$	Cu_2Se	2.67	$A_2^5B_5^6$	As_2Se_5	5.71
$A_2^2B^4$	Zn_2Ge	2.67	A^6	Se	6.00
A^3	Ga	3.00	$A^3B_3^7$	$GaBr_3$	6.00
$A_3^2B_2^5$	Zn_3As_2	3.20	$A^4B_4^7$	$GeBr_4$	6.40
$A_4^3B_3^4$	Ga_4Ge_3	3.43	$A_5^5B_5^7$	$AsBr_5$	6.67
A^4	Ge	4.00	$A_6^6B_6^7$	$SeBr_6$	6.86
A^1B^7	CuBr	4.00	A^7	Br	7.00

the analogue of substances of the type $A_3^1B^5$ (Cu_3As). It is noteworthy that the groups with 2·67 and 5·33 el/at* have two representatives each, which differ in their degree of ionic character in accordance with the position of their constituent elements in the periodic system.

As will be seen later, Table 9 can be used to select a series of types of binary compounds restricted by definite limiting values of the average number of electrons per atom. It may be assumed with a considerable degree of certainty that the representatives of these types will include substances with semiconducting properties and a structure related to that of zinc blende. It may be shown further that in addition to the diamond-like semiconductors, a number of structurally related substances form ternary and more complex analogues in accordance with the same laws which were obeyed in the formation of complex tetrahedral phases.

Let us examine the question as to what limiting values of the number of electrons per atom can be used to restrict the compounds of Table 9 in the search for semiconducting properties. In groups of elements with a number of electrons in excess of four, semiconducting properties are observed for various elements up to a number of electrons per atom equal to seven. Hence it may be assumed that semiconducting properties are also possible in their binary analogues up to the maximum number of electrons per atom.

On the other hand, elements having semiconducting properties and fewer than 3 el/at do not exist. However, semiconductors have been discovered among compounds of the type $A_3^1B^5$, for example, Li_3Bi and Na_3Sb [25, 26]. Therefore, semiconducting properties are possible in binary substances with a number of electrons per atom from 2 to 7. We shall now discuss these compounds from the standpoint of crystal chemistry.

Many of the substances shown in Table 9 may be considered to be related to the zinc blende group [5, 27, 28]. As a matter of fact, compounds of type $A_2^3B_3^6$ (Ga_2Se_3 in Table 9), which were mentioned above, crystallize in the structure of zinc blende with one-third of the unoccupied lattice points in the "cationic" part of the lattice. One can also imagine that compounds of the type $A^4B_2^6$ ($GeSe_2$ in Table 9) may crystallize in the structure of ZnS with one half of the vacant sites in the "cationic" part of the lattice, and so on up to the type $A^6B_6^7$ ($SeBr_6$). However, the problem of the actual existence of such binary compounds with the defect structure of zinc blende will depend primarily on the stability of the ZnS lattice with any given number of intrinsic defects.

---

* Number of electrons per atom.

Such a stability limit may be determined approximately by considering the structures of actually existing compounds which have a number of electrons greater than four. Whereas the tetragonal structure HgI_2 (type $A^2B_2^7$, 5·33 el/at) may still be considered related to zinc blende, it may be said that in compounds of type $A^4B_4^7$ such as SnI_4 (6·4 el/at) the number of defect sites in the cationic part is so great that no co-ordination lattices can form. Compounds of this type usually crystallize in molecular structures. As an arbitrary limit, one which is certainly too high, we shall take a number of electrons per atom intermediate between these values and equal to six, which corresponds to the $A^3B_3^7$ type of compound ($GaBr_3$).

The structures of substances with a number of electrons per atom less than 4 may be considered excess structures as compared to those of zinc blende, at least up to the type $A_3^1B^5$. In the ZnS structure, the metal atoms occupy half of the tetrahedral vacancies formed by the dense cubic packing of the non-metal.

In the structure of the type $A_3^2B_2^5 - Zn_3As_2$ (3·2 el/at), three-quarters, not one half of the tetrahedral vacancies, as in zinc blende, are occupied. The unit cell is tetragonal in this case only because of the ordering of the vacant and occupied tetrahedral sites.

In the structure of the type $A_2^1B^6$ and $A_2^2B^4$ (2.67 el/at), all the tetrahedral vacancies have already been occupied. Strictly speaking, when the number of electrons per atom is less than 2·67, structures crystallochemically close to zinc blende can no longer be formed. However, since the filling of octahedral vacancies does not rule out semiconducting properties, as, for example, in Li_3Bi [25] and Na_3Sb [26], this analogy may be extended to compounds of the type $A_3^1B^5$ with two electrons per atom. Thus, among binary compounds with a number of electrons per atom between 2 and 6, one can expect to find substances crystallizing in structures related to zinc blende and possessing the properties of semiconductors.

It may be assumed that each of these binary compounds has a number of ternary or more complex analogues formed according to the same rules which were followed in the formation of complex tetrahedral phases. Taking as the basis the rule of normal valence and the requirement that the average number of electrons per atom be equal to some particular value (from 2 to 6), one can show graphically and analytically all the possible types of analogue for each of the group of compounds related to zinc blende. Some of the ternary analogues have already been obtained; they crystallize in the predicted structure and are semiconductors.

As an example of the graph identification of types of ternary analogues, Fig. 8 shows triangular concentration diagrams of

1 – 4 – 5 and 2 – 2 – 6 on which are drawn the lines of normal valence and the isoelectronic lines corresponding to the number of electrons in binary compounds whose analogues we are seeking. Intersections of these lines give the ternary analogue compounds.

Let us examine a complete table of all the possible ternary analogues of binary compounds with a number of electrons per atom from 2 to 6 (Table 10).

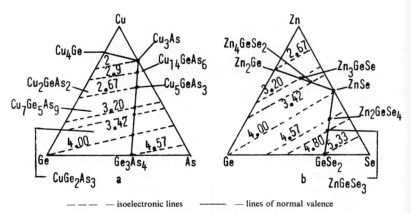

Fig. 8 *Diagrams of the formation of ternary analogues of binary defect and excess tetrahedral phases*

The probability of formation is not the same for all the types of compounds. In Table 10, the dotted lines frame the types of compounds which correspond to a monocationic combination, which, as we have seen in the example of tetrahedral phases, is an unfavourable factor for the formation of stable compounds. It is evident that we should completely exclude from these types of compounds those in which the "anions" consist of elements of groups II and III (such compounds are surrounded by double wavy lines). The types of compounds for which representatives are known are printed in bold type and framed; these will now be discussed.

Let us examine the compounds which are of excess type as compared with the structure of zinc blende. In compounds of the type $A_2^1B^2C^4$ (2 el/at), there exists Li_2MgSn (see column 1 – 2 – 4 of

Table 10), which crystallizes in the structure of DO_3 [27], i.e., in the structure of BiF_3, akin to ZnS, where, in contrast to the latter, all the tetrahedral and octahedral vacancies are occupied.

Semiconducting properties are probable for any representatives of this type, but thus far only one compound has been synthesized, and its electrical properties have not yet been studied.

For the structure of antifluorite with 2·67 el/at, a number of compounds of the type $A^1B^2C^5$ are known and were discussed above. The compounds $A_3^1B^3C_2^5$ and $A_3^1B^4C_3^5$ (columns 1 – 3 – 5 and 1 – 4 – 5 of Table 10), whose semiconducting properties have not been studied but are probable for certain of them, crystallize in the same structure and with the same number of electrons per atom [29-31].

Compounds of the type $A_5^2B_2^3$ (3·2 el/at) have no ternary analogues that have been studied, with the exception of compounds decomposable by water which have the structure of perovskite and correspond to the type $A_3^1B^6C^7$. Ag_3SBr and Ag_3SI have been obtained [32].

Compounds of the type $A_5^2B_2^3$ (Zn_5Ga_2, 2·29 el/at) and $A_4^3B_3^4$ (Ga_4Ge_3, 3·42 el/at) have no representatives crystallizing in a structure related to zinc blende. Their ternary analogues have not been studied either. However, the question of their existence may arise, since we shall consider below an example where ternary analogues were discovered in a similar case and were found to belong to the structure of zinc blende and to have semiconducting properties.

In the preceding section, we discussed compounds with 4 el/at. In Table 10, these compounds and their ternary analogues are the initial substances for the formation of defect and excess compounds with respect to the structure of zinc blende.

The first type of binary compound which may be supposed to crystallize in the structure of zinc blende with defects in the cationic region are compounds of the type $A_3^4B_4^5$ (Ga_3As_4, 4·57 el/at). The binary compounds with this formula which actually exist, Si_3N_4 and Ge_3N_4, crystallize in another structure (which contains a three-dimensional arrangement of the tetrahedra). Nevertheless, ternary analogues of this type include the compounds $A_2^1B^2C_4^1$ (column 1 – 2 – 7 of Table 10), whose representatives, Cu_2HgI_4 and Ag_2HgI_4, indeed crystallize in a structure which differs from zinc blende only in the defects of the cationic part[33]. The semiconducting properties of these compounds have not yet been studied.

The substances $ZnGa_2Se_4$, $CdIn_2Te_4$, $CdIn_2Se_4$, etc. (see column 2 – 3 – 6 of Table 10), which are analogues of $A_3^4B_4^5$, crystallize in a structure similar to zinc blende[6]. Their semiconducting properties have been known for a long time and have been studied rather

Table 10 Ternary analogues of ...

Formula	Number of el/at	1—2—3	1—2—4	1—2—5	1—
$A_3^1B^5$ (Cu$_3$As)	2.0	$A^1B_4^2C^3$	$A_2^1B^2C^4$	$A_3^1B^5$	A_{12}^1
$A_5^2B_2^3$ (Zn$_5$Ga$_2$)	2.29	$A_5^2B_2^3$	$A_2^1B_3^2C_2^4$	$A_4^1B^2C_2^5$	A_{20}^1
$A_2^1B^6$; $A_2^2B^4$ (Cu$_2$Se; Zn$_2$Ge)	2.67	—	$A_2^2B^4$	$A^1B^2C^5$	A
$A_3^2B_2^5$ (Zn$_3$As$_2$)	3.20	—	—	$A_2^3B_2^5$	A_2^1
$A_4^3B_3^4$ (Ga$_4$Ge$_3$)	3.42	—	—	—	A_2^1
A^3B^5; A^2B^6; A^1B^7 (GaAs; ZnSe; CuBr)	4.00	—	—	—	A
$A_3^4B_4^5$ (Ge$_3$As$_4$)	4.57	—	—	—	—
$A_2^3B_3^6$ (Ga$_2$Se$_3$)	4.80	—	—	—	—
$A^2B_2^7$; $A^4B_2^6$ (ZnBr$_2$; GeSe$_2$)	5.33	—	—	—	—
$A_2^5B_5^6$ (As$_2$Se$_5$)	5.71	—	—	—	—
$A^3B_3^7$ (GaBr$_3$)	6.00	—	—	—	—

Formula	Number of el/at	1—5—6	1—5—7	1—6—7	2—
$A_3^1B^5$ (Cu$_3$As)	2.0	$A_3^1B^5$	$A_3^1B^5$	—	
$A_5^2B_2^3$ (Zn$_5$Ga$_2$)	2.29	$A_5^1B^5C^6$	$A_{10}^1B_3^5C^7$	—	A
$A_2^1B^6$; $A_2^2B^4$ (Cu$_2$Se; Zn$_2$Ge)	2.67	$A_2^1B^6$	$A_4^1B^5C^7$	$A_2^1B^6$	A
$A_3^2B_2^5$ (Zn$_3$As$_2$)	3.20	$A_{11}^1B^5C_8^6$	$A_6^1B^5C_3^7$	$A_3^1B^6C^7$	A^2

ounds related to zinc blende*

—7	1—3—4	1—3—5	1—3—6	1—3—7	1—4—5	1—4—6	1—4—7
$_2^2C_3^7$	$A_5^1B^3C_2^4$	$A_3^1B^5$	$A_9^1B^3C_2^6$	$A_6^1B^3C^7$	$A_3^1B^5$	$A_6^1B^4C^6$	$A_9^1B_2^4C^7$
$_3^2C_7^7$	$A_7^1B_3^3C_4^4$	$A_9^1B^3C_4^5$	$A_{15}^1B^3C_5^6$	$A_{20}^1B_3^3C_5^7$	$A_{14}^1B^4C_6^5$	$A_{10}^1B^4C_3^6$	$A_5^1B^4C^7$
$^2C_4^7$	$A^1B^3C^4$	$\mathbf{A_3^1B^3C_2^5}$	$A_2^1B^6$	$A_8^1B^3C_3^7$	$\mathbf{A_5^1B^4C_3^5}$	$A_2^1B^6$	$A_6^1B^4C_2^7$
$^2C_9^7$	$A^1B_5^3C_4^4$	$A_3^1B_3^3C_4^5$	$A_5^1B^3C_4^6$	$A_{12}^1B^3C_7^7$	$A_2^1B^4C_2^5$	$A_8^1B^4C_6^6$	$A_9^1B^4C_5^7$
$^2C_{14}^7$	$A_4^3B_3^4$	$A_3^1B_5^3C_6^5$	$A_3^1B^3C_3^6$	$A_{16}^1B^3C_{11}^7$	$A_7^1B_5^4C_9^5$	$A_{10}^1B_2^4C_9^6$	$A_{12}^1B^4C_8^7$
B^7	—	A^3B^5	$\mathbf{A^1B^3C_2^6}$	A^1B^7	$\mathbf{A^1B_2^4C_3^5}$	$\mathbf{A_2^1B^4C_3^6}$	A^1B^7
$^2C_4^7$	—	—	$A^1B_3^3C_8^6$	$A_5^1B^3C_8^7$	$A_3^4B_4^5$	$A_4^1B_5^4C_{12}^6$	$A_8^1B^4C_{12}^7$
$^2C_3^7$	—	—	$A_2^3B_3^6$	$A_3^1B^3C_6^7$	—	$A_2^1B_4^4C_9^6$	$A_5^1B^4C_9^7$
B_2^7	—	—	—	$A^1B^3C_4^7$	—	$A^4B_2^6$	$A_3^1B^4C_6^7$
-	—	—	—	$A^1B_3^3C_{10}^7$	—	—	$A^1B^4C_5^7$
-	—	—	—	$A^3B_3^7$	—	—	$A^1B_2^4C_9^7$

—5	2—3—6	2—3—7	2—4—5	2—4—6	2—4—7	2—5—6	2—5—7
-	—	—	—	—	—	—	—
B_2^3	$A_5^2B_2^3$	$A_5^2B_2^3$	—	—	—	—	—
$^3C^5$	$A_6^2B_2^3C^6$	$A_3^2B_3^3C^7$	$A_2^2B^4$	$A_2^2B^4$	$A_2^2B^4$	—	—
B_2^5	$A_9^2B_2^3C_4^6$	$A_3^2B^3C^7$	$A_3^2B_2^5$	$A_3^2B^4C^6$	$A_9^2B_4^4C_2^7$	$A_3^2B_2^5$	$A_3^2B_2^5$

Formula	Number of el/at	1—5—6	1—5—7	1—6—7	2—3
$A_4^3B_3^4$ (Ga_4Ge_3)	3.42	$A_7^1B^5C_6^6$	$A_8^1B^5C_5^7$	$A_4^1B^6C_2^7$	A_4^3
A^3B^5; A^2B^6; A^1B^7 (GaAs; ZnSe; CuBr)	4.00	$A_3^1B^5C_4^6$	A^1B^7	A^1B^7	—
$A_3^4B_4^5$ (Ge_3As_4)	4.57	$A_7^1B_5^5C_{16}^6$	$A_{11}^1B^5C_{16}^7$	$A_{14}^1B^6C_{20}^7$	
$A_2^3B_3^6$ (Ga_2Se_3)	4.80	$A^1B^5C_3^6$	$A_7^1B^5C_{12}^7$	$A_9^1B^6C_{15}^7$	—
$A^2B_2^7$; $A^4B_2^6$ ($ZnBr_2$; $GeSe_2$)	5.33	$A^1B_3^5C_8^6$	$A_3^1B^5C_8^7$	$A_4^1B^6C_{10}^7$	—
$A_2^5B_5^6$ (As_2Se_5)	5.71	$A_2^5B_5^6$	$A_5^1B_3^5C_{20}^7$	$A_7^1B_3^6C_{25}^7$	—
$A^3B_3^7$ ($GaBr_3$)	6.00	—	$A^1B^5C_6^7$	$A_3^1B_2^6C_{15}^7$	—

Formula	Number of el/at	2—6—7	3—4—5	3—4—6	3—4
$A_3^1B^5$ (Cu_3As)	2.0	—	—	—	—
$A_5^2B_2^3$ (Zn_5Ga_2)	2.29	—	—	—	—
$A_2^1B^6$; $A_2^2B^4$ (Cu_2Se; Zn_2Ge)	2.67	—	—	—	—
$A_3^2B_2^5$ (Zn_3As_2)	3.20	—	—	—	—
$A_4^3B_3^4$ (Ga_4Ge_3)	3.42	—	$A_4^3B_3^4$	$A_4^3B_3^4$	A_4^3
A^3B^5; A^2B^6; A^1B^7 (GaAs; ZnSe; CuBr)	4.00	A^2B^6	A^3B^5	$A_2^3B^4C^6$	A_3^3B
$A_3^4B_4^5$ (Ge_3As_4)	4.57	$A_3^2B_2^6C_2^7$	$A_3^4B^5$	$A_6^3B^4C_7^6$	A_9^3B
$A_2^3B_3^6$ (Ga_2Se_3)	4.80	$A_2^2B^6C_2^7$	—	$A_2^3B_3^6$	A_2^3B
$A^2B_2^7$; $A^4B_2^6$ ($ZnBr_2$; $GeSe_2$)	5.33	$A^2B_2^7$	—	$A^4B_2^6$	A_3^3B

3—5	2—3—6	2—3—7	2—4—5	2—4—6	2—4—7	2—5—6	2—5—7
$B_3^3C_3^5$	$A_{12}^2B_2^3C_7^6$	$A_{16}^2B_5^3C_7^7$	$A_7^2B^4C_6^5$	$A_4^2B^4C_2^6$	$A_{12}^2B_5^4C_4^7$	$A_4^2B_2^5C^6$	$A_8^2B_5^5C^7$
$^3B^5$	A^2B^6	$A_4^2B^3C_3^7$	$\mathbf{A^2B^4C_2^5}$	A^2B^6	$A_3^2B^4C_2^7$	A^2B^6	$A_2^2B^5C^7$
—	$\mathbf{A^2B_2^3C_4^6}$	$A_6^2B^3C_7^7$	$A_3^4B_4^5$	$\mathbf{A_2^2B^4C_4^6}$	$A_9^2B_2^4C_{10}^7$	$A_7^2B_2^5C_{12}^6$	$A_3^2B^5C_3^7$
—	$A_2^3B_3^6$	$A_8^2B^3C_{11}^7$	—	$\mathbf{A^2B^4C_3^6}$	$A_6^2B^4C_8^7$	$A_4^2B_2^5C_9^6$	$A_4^2B^5C_5^7$
—	—	$A^2B_2^7$	—	$A^4B_2^6$	$A^2B_2^7$	$A^2B_2^5C_6^6$	$A^2B_2^7$
—	—	$A^2B^3C_5^7$	—	—	$A_3^2B^4C_{10}^7$	$A_2^5B^6$	$A_5^2B^5C_{15}^7$
—	—	$A^3B_3^7$	—	—	$A^2B^4C_6^7$	—	$A_2^2B^5C_9^7$

5—6	3—5—7	3—6—7	4—5—6	4—5—7	4—6—7	5—6—7	
—	—	—	—	—	—	—	
—	—	—	—	—	—	—	
—	—	—	—	—	—	—	
—	—	—	—	—	—	—	
—	—	—	—	—	—	—	
$^3B^5$	A^3B^5	—	—	—	—	—	
$B^5C_3^6$	$A_6^3B_5^5C_3^7$	—	$A_3^4B_4^5$	$A_3^4B_4^5$	—	—	
$A_2^3B^6$	$A_4^3B_3^5C_3^7$	$A_2^3B_3^6$	$A_2^4B_2^5C^6$	$A_4^4B_5^5C^7$	—	—	
$B^5C_4^6$	$A_2^3B^5C_3^7$	$A^3B^6C^7$	$A^4B_2^6$	$A^4B^5C^7$	$A^4B_2^6$	—	

Formula	Number of el/at	2—6—7	3—4—5	3—4—6	3—4—7
$A_2^5 B_5^6$ (As$_2$Se$_5$)	5.71	$A_7^2 B^6 C_{20}^7$	—	—	$A_6^3 B^4 C_{14}^7$
$A^3 B_3^7$ (GaBr$_3$)	6.00	$A_3^2 B^6 C_{12}^7$	—	—	$A^3 B_3^7$

*Asterisk denotes substances crystallizing in the spinel structure; the remaining substances crystallize in tetrahedral structures. In the compound ZnAl$_2$S*, the tetrahedral (wurtzite) structure appears only above 1000°C

thoroughly[34]. Analogues of $A_3^4 B_4^5$ also include the substances $A_2^2 B^4 C_4^6$ (column 1 – 2 – 7 of Table 10). These substances (Zn$_2$GeS$_4$, Zn$_2$GeSe$_4$) have been prepared recently, and their structure was found to be diamond-like; they are phases of variable composition[35].

Finally, the same group of ternary substances which are analogues of $A_3^4 B_4^5$ includes a large number of phases of variable composition based on the compounds $A^3 B^5 - A_2^3 B_3^6$. The substances corresponding in composition to the intermediate points of this system are $A_3^3 B^5 C_3^6$ (column 3 – 5 – 6 of Table 10) which are ternary analogues of $A_3^4 B_4^5$ with 4·57 el/at. In contrast to all the preceding ternary analogues, these substances are monocationic but nevertheless stable. The cause of this is unclear, but it may be significant that the vacant sites in the structure appear to play the part of cations of the second kind, and from the standpoint of this interpretation we are dealing with dicationic quaternary alloys, not with monocationic ones. Incidentally, these quaternary alloys are, as a rule, phases of variable composition as are the phases of the type $A_3^3 M^5 C_3^6$.

The existing ternary analogues of compounds of the type $A_2^3 B_3^6$ are the phases of variable composition ZnGeS$_3$ and ZnGeSe$_3$ (column 2 – 4 – 6 of Table 10), which have a region of homogeneity in the section ZnS – GeS$_2$ and ZnSe – GeSe$_2$ (see the discussion of $A_2^2 B^4 C_4^6$ above[35]).

The existing ternary compounds with the formula $A^3 B^5 C_4^6$ (ana-

Table 10 cont.

3—5—6	3—5—7	3—6—7	4—5—6	4—5—7	4—6—7	5—6—7
$A_2^5B_5^6$	$A_4^3B^5C_9^7$	$A_2^3B^6C_4^7$	—	$A_4^4B_3^5C_7^7$	—	$A_2^5B_5^6$
—	$A^3B_3^7$	$A^3B_3^7$	—	$A_2^4B^5C_5^7$	$A^4B^6C_2^7$	$A^5B_2^6C^7$

The dashes indicate that in these combinations there is no tetrahedral phase

logues of $A^4B_2^6$) include a series of compounds where A^3 is boron or aluminium, B^5 is phosphorus and arsenic, and C^6 is sulphur, selenium or tellurium. Not one of these compounds crystallizes in the diamond-like structure; most of them are similar in structures to SiS_2[36]. However, this structure also involves a tetrahedral arrangement of atoms, but the tetrahedra form chains.

In comparing the existing excess ternary analogues with the defect ones, our attention is drawn to the fact that the latter are in most cases phases of variable composition.

To date, only a small number of compounds of the types listed in Table 10 has been studied; this applies particularly to the "excess" phases, where syntheses were usually limited to combinations of elements of the first periods. It is characteristic that the composition of the "excess" phases includes not only the elements participating in the structure of the usual defect tetrahedral phases, but also others such as lithium, sodium or bismuth. In this respect and also in others (small capacity to form phases of variable composition, etc.) the "excess" phases stand somewhat alone, differing from normal and defect tetrahedral phases.

Following the rules for building up analogues corresponding to those given in Section 2, we can derive formulae for quaternary and more complex defect and excess tetrahedral phases. This is not necessary, however, since they will be combinations of binary and ternary phases, a complete list of which is given in Table 10. It is evident that Table 10 may be used to select the direction of the search for new semiconducting materials with structures akin to that of zinc blende.

The proposed method of determining the number of possible complex analogues is rather general, as can be shown by examining the conditions associated with the building up of other crystallochemical groups of semiconductors. Let us try to relate these conditions to the principle of building up the groups of the Periodic System. (When applied to crystallochemical groups of unsaturated valence and mixed phases, these conditions may be more complex.)

It is known that the factor determining the similarity of elements of the same group and the difference between elements of different groups of the Periodic System is the number of electrons of the outermost shell of the atom, or the concentration of the valence electrons. (In Section 2 of this chapter, we have already used this condition for building up one of the crystallochemical groups, the group of diamond-like semiconductors.)

Thus far, the electron concentration was considered to be a factor restricting the phase composition only for metallic phases in which the relative proportions of the components vary between wide limits. In our view, this factor should also be the determining one (in the sense that it determines the category to which the phase belongs) both for chemical compounds and for complex phases.

It may be assumed that the first requirement for the formation, and hence for the classification, of semiconductor phases should be a given average number of valence electrons per atom. The second requirement follows logically from the law governing the building up of the Periodic System of elements.

As we know, the similarity between elements of the same group manifests itself only when they achieve a valence of the same type. Complex substances may belong to the same category if the elements constituting them achieve the same type of valence (normal valence, partial valence, etc.). These requirements may be illustrated by the example of the family of tetrahedral phases, which have been studied most thoroughly.

As seen from Table 10, all the hypothetical and actually existing substances given there satisfy the requirement of a given electron concentration.

The second requirement is also satisfied: the number of electrons given up for the formation of bonds by atoms occupying the usual sites of "cations" in the crystal lattice is equal to the number of electrons missing in the outermost shells of the atoms occupying the

FORMATION OF DIAMOND-LIKE SUBSTANCES

sites of anions.

To put it another way, the atoms which are "cations" achieve normal valence with respect to hydrogen, those which are "anions", with respect to oxygen; both of these requirements define a category of substances in which there may exist crystallochemical groups or families.

A similar situation prevails in group IVb of the Periodic System, where the subgroup of C, Si, Ge, Sn and Pb contains the crystallochemical group of diamond, Si, Ge and grey tin. The common characteristic of the phases of this crystallochemical group or family is the similarity of their short-range order.

The crystallochemical characteristics and similarities of the phases are determined not by the symmetry of the arrangement of the atoms (lattice symmetry), but by the similarity in the short-range ordering of the atoms, and it is this feature which actually determines a crystallochemical group. This may be illustrated by taking as the reference the close packing of non-metal atoms which are "anions".*

The short-range order is practically the same in all types of close packing; the types differ only in the manner in which the atomic layers are superimposed.

The immediate vicinity of a metal atom ("cation") located in a tetrahedral or octahedral site is exactly the same in any close packing.

The immediate vicinity of a non-metal atom ("anion") occupying in the crystal structure a position creating a close packing may vary (1) because of changes in the relative distribution of the octahedral and tetrahedral vacancies (the layered character of the structure) and (2) because of changes in the number of occupied octahedral and tetrahedral vacancies. The latter is the cause of the formation of defect and excess structures.

Experiments have shown that almost all of the substances in this group which have been studied possess semiconducting properties. Regularities in the variation of the basic physicochemical and semiconducting properties of the substances in this family have also been pointed out.

Another family of semiconducting phases (family or group of analogues of PbS) may be formed in similar fashion. In this case, an incomplete valence of the metal atoms is achieved. This is due to the stability of the s^2 shells of the elements in the lower periods. The electron concentration (first requirement) also corresponds to

---

* Strictly speaking, the terminology employed here is applicable only to ionic compounds, but we are using it because the arrangement of the atoms in the interesting structures of the semiconducting phases is the same as in ionic compounds.

the specified conditions, but its calculation takes into account only those valence electrons which participate in the formation of the bond.

It is possible to calculate the composition of the likely binary and ternary substances in which the electron concentration varies from 2 to 6 el/at and an incomplete valence of the metal atoms (smaller by 2 units) is achieved. The crystallochemical picture becomes considerably more complicated in this case. The formation of a family of mixed phases is also possible; this family should include substances in which a part of the atoms achieve complete valence and the other part an incomplete valence. All these families are represented primarily by compounds of elements in b subgroups of the Periodic System (classification according to reference 48). The likelihood of the participation of any given elements is determined by different conditions for each family or group.

Elements with unfilled d shells often form semiconducting phases, but the role of the d electrons and their participation in the formation of chemical bonds are quite complex. It is therefore reasonable to classify the semiconducting phases involving transition elements in a separate category. Substances in this category undoubtedly have special properties which they alone possess. All that can be said definitely is that a more thorough investigation is required and development will be based on the accumulation and analysis of experimental data on new semiconductors.

Lately, attempts have been made to formulate rules or criteria to be followed in the search for new semiconducting substances. Thus, two particularly useful criteria have been noted in the work of Goodman[16]:

1. The possibility of representing the formula of the compound in the ionic form.

2. The existence of the compound in the form of a linear phase with the maximum melting point on the corresponding phase diagram.

Certain other criteria are also given: brittleness, high resistivity, etc.

Mooser and Pearson[27] suggest four conditions for the achievement of a "semiconducting bond" which in their opinion determines the semiconducting properties of a substance.

1. The bonds should be primarily covalent in character.

2. The bonds are formed by the pairing of electrons, which leads to the formation of completely filled s and p subshells of all the atoms in elemental semiconductors. In the case of semiconducting compounds, it is necessary that the subshell of one of the two atoms

be filled.

3. The presence of free "metallic" orbits in some of the atoms of the compound does not impair its semiconducting properties, provided these atoms are not directly linked with one another.

4. The bonds form a continuous one-, two- or three-dimensional network which encompasses the entire crystal.

These and similar criteria are useful, although not exhaustive, as are all the concepts presented above. The basis for a further elaboration of the principles and criteria of formation of semiconducting phases should be the Periodic System of the elements and the laws governing its construction.

CHAPTER 2

PHYSICOCHEMICAL AND ELECTRIC PROPERTIES OF DIAMOND-LIKE SEMICONDUCTORS

Part A

Elemental Semiconductors

INTRODUCTION

We shall begin the description of the physicochemical and electric properties of diamond-like semiconductors with the elements: diamond, silicon, germanium and grey tin. Silicon carbide and solid solutions of silicon and germanium, for which brief data will be given, are similar in properties to this group of elemental semiconductors.

Since germanium and, in recent years, silicon, have found very broad applications in radio electronics, there is at the present time a large number of papers in the literature devoted to the preparation of these substances in the pure state, to the study of their semiconducting properties, etc. Special collections have been published, devoted to these semiconductors[37-42], containing data on the electric, thermoelectric, galvanomagnetic, optical and other properties of semiconducting elements. We shall, therefore, give below only the data for the basic physicochemical and electric properties of the semiconductors of this group. The problems of preparation and purification will be touched upon very briefly.

The basic characteristics of semiconductors are their crystal structure and the lattice spacings, knowledge of which provides first insight into short-range order.

Short-range order plays a decisive part in the electrical properties of a solid, and by short-range order, as was indicated above are meant the chemical nature of the atoms, the geometry of their arrangement, and the absolute distance between them. The two latter parameters are determined in turn by the nature of the chemical bond between the atoms. Within the same type of crystal structure, the character of the change in the chemical bonding depends on the change in the nature of the atoms – primarily on the change in atomic weight. As this changes, so do the lattice spacings, which are important characteristics of a substance.

The identification of the types of bonding present in a given semiconductor is a different matter.

Lately, cleavage, defined as the splitting of crystals along any crystallographic direction, has been used to evaluate ionic and covalent types of bonding in semiconductors.

At the present time, there are no other simple and reliable experimental methods* permitting such an evaluation, and data on cleavage are therefore very useful. This applies particularly to binary compounds, which are described in the next section. There are, however, other experimental approaches which show promise.

A. F. Ioffe wrote the following about the mechanical properties of semiconductors[43]: "The basic properties of a semiconductor are determined by the bonding forces operating in it, which are manifested primarily in the mechanical strength and other elastic properties".

In the case of diamond-like semiconductors, one such basic property is hardness. The measurement of hardness was introduced into physicochemical analysis by N. S. Kurnakov, and proved to be one of the most sensitive methods for the study of solids. The experimental technique successfully developed for alloys was found to be suitable for semiconductors. Microhardness, usually determined by indentation with a square diamond pyramid (or Knoop pyramid) was found to be a reproducible characteristic of semiconductors. The method for measuring microhardness has been found to be useful not only in the study and identification of the structural components of semiconducting materials, but also in evaluating the correlation between type of bonding and crystallographic direction (not only within certain crystallochemical groups, but between them as well).

The thermodynamics of semiconductors has not been sufficiently developed to date. However, in view of the great practical and

---

* Lately, studies have appeared dealing with the determination of the effective charge of atoms, and are being used for the same purpose [580 and others.]

theoretical importance of the problems pertaining to this area (explanation of the conditions necessary for various processes, calculation of bond energy, etc.) it is necessary to consider the data available in the literature on the thermodynamic characteristics of substances.

Among the chemical properties of diamond-like semiconductors, those of prime importance are the data on the solubility and stability in relation to the action of various agents, particularly oxygen and the humidity of the air.

In every area of technological application, there arises a number of specific requirements with respect to the basic characteristics of semiconducting materials. Since the area of application of diamond-like semiconductors is chiefly radio electronics, this book cites some of the electrical parameters which play a decisive part in the choice of semiconducting materials in this field*. An example of such a parameter is the forbidden energy gap width (the activation energy of intrinsic conduction), expressed in electron volts.

In order to provide conduction, a relatively pure semiconductor whose impurities are already ionized at a given temperature must contribute some of its intrinsic electrons to the process; these are electrons which form the chemical bonds between the atoms of the semiconducting substance. The electrons must, therefore, be excited by an energy which is specific for each semiconductor, is referred to as the activation energy of intrinsic conduction, and corresponds in magnitude to the forbidden gap width. The latter term is related to the energy spectrum of the semiconductor, where two energy bands are recognized: a band which is completely filled at low temperature, i.e., the valence band, and an allowed band following it, i.e., the conduction band. The energy difference between them is the forbidden gap width. This quantity undoubtedly depends on the energy of the chemical bond and its characteristics.

The mobility of the charge carriers, a second important parameter of a semiconductor, characterizes the motion of electrons in an electric field, and is determined by the mechanism of scattering of the carriers. In semiconductors, the carriers are scattered both by the thermal vibrations of the lattice and by the impurities. Thus, the magnitude of the mobility, roughly speaking, is a measure of the purity of the material in those cases where the maximum mobility has already been obtained experimentally from other samples of this substance, or when it is known from theory. The mobility of the carriers is measured in $cm^2/Vsec$.

---

* For more detail on the physical parameters and methods of their measurement see references 44, 45.

The operation of most of the semiconductor devices is governed by the impurity mechanism of conduction, whereas the intrinsic conduction, which is associated with a definite activation energy of the electrons from their parent atoms in the semiconductor (i.e., the forbidden gap width), usually determines the permissible temperature at which the devices can be used.

The function of the impurities may be fulfilled not only by foreign atoms, but also by defects in the crystal lattice. At room temperature, practically all the impurities are ionized and create a definite concentration of impurity carriers denoted by n and measured in cm^{-3}.

For an "ideal" semiconductor of the type of diamond, the behaviour of foreign impurities which form a substitutional-type solid solution is very simple. If the impurity atom has a higher valence than the atom of the basic lattice, the excess valence electron of the impurity atom may take part in the transfer of electricity, giving rise to the electronic conduction of the substance.

If the impurity atom has a lower valence, it cannot complete the structure of valence bonds surround it, and a "hole" is formed in the system of bonds. The displacement of this hole by its filling with electrons from the adjacent bonds gives rise to hole conduction. For impurities penetrating into the interstices, the mechanism is more complicated.

Of great importance for the practical use of semiconductors of the diamond crystallochemical group is the possibility of forming a transition layer, the so-called *p-n* junction (from "positive" and "negative"), where regions of conductivities of different signs – the hole conductivity and the electronic conductivity – are either in contact or penetrate each other. The physical bases of most of the technical applications of diamond-like semiconductors are the electronic processes taking place in this transition region. It is thus easy to understand the importance assumed by the development of methods of purification, alloying and preparation of perfect single crystals of semiconductors with reproducible properties; there is extensive literature devoted to these problems[37, 38, 40]. We will be citing here only the basic data pertaining to the effect of impurities on the conduction type of substance, and data on the distribution coefficients of the impurities.

This last characteristic is important because the generally accepted metallurgical methods of purification and preparation of silicon and germanium single crystals are based, in their last stage, on the use of the difference in the composition of the adjacent solid and liquid phases of the substance during crystallization, and on the forcing of the impurities away from the crystallization front[37–39]. The boundary between the solid and the liquid phase of the substance on which the crystal is growing is displaced. However, this displacement occurs at such a slow rate that the conditions at the crystallization front are close to equilibrium conditions. The process can therefore be studied by means of fusibility diagrams of two-component systems, where one component is the semiconductor and the other is the impurity.

The liquidus and solidus curves in the region of low impurity concentrations may be represented with sufficient accuracy in the form of straight lines intersecting at the melting point of the basic semiconductor component. Two cases are possible, depending on whether the impurity lowers or raises the melting point of the basic semiconductor, and are illustrated in Fig. 9, in which impurity concentration, C, is plotted against temperature, T. In the first case (Fig. 9a) the ratio of the impurity concentration in the solid phase to the impurity concentration in the adjacent liquid, given by the distribution coefficient $K = c_f/c_\rho$, will be less than unity. In the second case (Fig. 9b), the distribution coefficient will be greater than unity. When K is close to unity, practically no purification takes place. The distribution coefficients of the impurities are used to evaluate the suitability of the metallurgical methods of purification.

The purified material must be given very definite electrical and physical properties. This is done by doping, i.e. introducing special impurities into the material. The technology of this process is also quite advanced. Here it will only be mentioned and brief data will also be given on the structural defects (dislocations) in silicon and germanium and on methods of revealing them.

It was stated above that the impurities are ionized in semiconductors at room temperature. This is because the removal of the electrons of the impurity atoms is facilitated by the polarizability of the atomic medium into which the impurity atom is introduced. The polarizability of the medium, characterized by the dielectric constant, relaxes the bonding forces between the electrons and the nucleus of the impurity atom and decreases the ionization energy. The dielectric constant ε is related to the activation energy of the intrinsic electrons E (for certain semiconductors their product $\varepsilon E =$ const). Thus, this characteristic of semiconductors is also important in judging their nature.

The optical properties of semiconductors, for instance, the position of the longwave edge in the absorption spectrum, can be used to determine the forbidden gap width in appropriate cases. Many of the values of this parameter given below were determined in this fashion. The sensitivity maximum in the spectral distribution of the photoelectric effect arises at wavelengths of incident light approximately corresponding to the absorption band. Thus, this parameter also enables one to evaluate the fundamental properties of the semiconductor and will be referred to later.

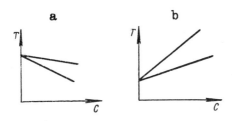

(a) the impurity lowers the melting point
(b) the impurity raises the melting point

Fig. 9 *Initial portion of the fusibility diagram*

Since in examining the semiconductors of the diamond group we have cited the characteristics which determine the choice of materials for semiconductor electronics, some mention should also

Table 11 *Requirements for semiconducting materials*

Characteristic of semiconducting materials	Requirements desired	Permissible impurity content (per atom)
Width of forbidden energy gap	> 0.7 eV	$\sim 10^{-4}$
Carrier mobility	$> 1000 \dfrac{cm^2}{V\ sec}$	$\sim 10^{-6}$
Conductivity	Controllable magnitude and type of conduction	$\sim 10^{-8}$
Lifetime of charge carriers	$> 10^{-5}$ sec	$\sim 10^{-10}$
Property of surface	Stability	?

be made of the lifetime of the charge carriers, which is a time constant characterizing the process of recombination of excess carriers. As we know, it is precisely the possibility of creating excess carriers in diamond-like semiconductors that accounts for the use of this group of materials in numerous devices, and the lifetime of the current carriers is of importance in determining their behaviour.

Table 11 lists some tentative requirements with respect to the characteristics of materials used in the preparation of some important types of devices in present semiconductor electronics[39]. It should be noted that for many types of devices these requirements will be different.

1. Diamond

Diamond is a transparent, hard, refractory crystal which has long been known as a precious stone. Its crystal structure has given its name to a structural type. The number of atoms in the unit cell of diamond is equal to eight. Half the atoms are at the corners of a face-centred cubic lattice, the other half occupy the centres of four of the eight octants. The octants are formed by three mutually perpendicular planes passing through the centre of the cell parallel to its edges. In the structure of diamond, every atom surrounded by four others forms four covalent bonds by giving up one of its valence electrons for use in each bond and by accepting a second electron from the neighbouring atom. This process of formation of the chemical bond in diamond as well as its analogues is described in quantum chemistry as follows:

An isolated carbon atom has 4 electrons with a principal quantum number of 2. As a result of the perturbing effect of the neighbouring atoms, one of the $2s$ electrons of the atom migrates to the p cell of the phase space. This transfer is energetically advantageous, since stronger bonds are formed with p electrons.

Furthermore, a linear combination of the wave eigen functions of $1s$ and $3p$ electrons, which leads to the formation of four identical intermediate (hybrid) electron states, is energetically advantageous. The clouds associated with each of these four electrons overlap in the tetrahedral directions with the clouds of the neighbouring atoms, and thus a saturated, strong, paired-electron and spherically unsymmetrical bond is formed.

Investigations of the spatial distribution of the electron density in diamond have shown that the electron density in the tetrahedral directions does not drop to zero as in the case of ionic compounds, but forms electron "bridges"[46, 47].

Space group of diamond $Fd3m$
Lattice period[48]   3·5597 Å
Hardness on Moh's scale   10

Diamond is the hardest of all minerals. In reference 49, the microhardness of diamond was measured with a Knoop diamond indenter. After seven measurements the indenter broke. Nevertheless, a value was obtained for the microhardness, $H_{100} = 8820 \pm 1380$ kg/mm^2.

Diamond has a well-defined cleavage along the octahedron (111) and also a microcleavage along the planes between (111) and (110) in the zone (110). The formation of such planes requires the rupture of the smallest possible number of bonds, and for this reason the cleavage corresponds to the ideal covalent bond and an undistorted surface[50].

Specific gravity   3·51 g/cm^2
Heat of sublimation   171·7 kcal/mole
Coefficient of linear expansion   1·18 x 10^{-6} deg^{-1}
Index of refraction   2·42
Dielectric constant[44, 51]   5·7

Several types of diamonds exist in nature which differ in the degree of perfection of the crystals and their physical properties. A survey of the electrical properties of diamond is given in references 52, 53.

Reference 54 deals with the preparation of artificial semiconducting diamonds. The problem of the synthesis of diamonds is also discussed in reference 55. The difficulties involved in the preparation are apparently due to the fact that diamond is a more stable modification than graphite only at very high pressures. Diamond is almost an insulator in its electric properties.

Forbidden gap with ΔE (from optical data)   5·6 eV
Electron mobility μ_n (at room temperature) $\sim$ 1800 cm^2/Vsec
Hole mobility μ_p $\sim$ 1200 cm^2/Vsec

The practical application of diamond in electronics is due to its ability to increase its electrical conductivity considerably when it is exposed to nuclear particles. This makes it possible to use diamonds as crystal counters operating at room temperature[41].

2. Graphite

Graphite, the hexagonal crystalline modification of carbon, is a black substance, greasy to the touch. The atoms of carbon form hexagonal networks in which the distances between the atoms are considerably less than the distances between the layers, i.e., graphite has a layered structure. The layers are oriented with respect to one another in such a way that the vacant centres of the hexagons of each layer are located above the filled corners of the adjacent layer. The cleavage of graphite is well defined and corresponds to the direction parallel to the layers.

Space group   C6/mmc
Lattice spacings[48] $a = 2·460$ Å, $c = 6·709$ Å
Specific gravity   2·22 g/cm^3

It is assumed that the atoms inside the layers are bound by covalent forces of considerable metallic bond character. The bond between the layers is due to van der Waals' forces.

Graphite is extracted from natural deposits but may be obtained artificially by heating anthracite in electric furnaces at 2500 – 3000 °C.

Graphite absorbs certain substances (fluorine, oxygen, etc.) which become embedded in the space between the layers of (111).

Studies of the electric properties of graphite single crystals have shown that the conductivity along the plane of the crystal is 100 times as great as the conductivity along the normal to this plane[42]. It is in this latter direction that monocrystalline graphite possesses semiconducting properties[44].

3. Silicon

Silicon is a dark grey substance with a tarry lustre. It crystallizes in the structure of diamond. The cleavage differs somewhat from the cleavage of diamond[50]. In addition to the good cleavage along (111) there is also observed a microcleavage along (110) and between (111) and (001) in the zone (11⁰).

Lattice spacing[37]   5·4198 Å
Moh's hardness   7
Microhardness (Knoop)[49] H_{25}

Table 12 lists certain properties of silicon and germanium.

Crystalline silicon is chemically more inert than the amorphous silicon usually obtained by chemical reactions. Silicon is stable in air and the kinetics of oxidation of silicon and germanium at room temperature have been studied[59,60].

Silicon is soluble in many fused metals e.g., zinc, aluminium, tin, lead, etc. It forms a continuous series of solid solutions only with germanium. As a rule the solubility of a metal in solid silicon is low; this is the basis for one of the methods of obtaining silicon and germanium in the form of single crystals, they being crystallized (with slow cooling) from a saturated solution in indium or gallium[38(p.45)].

At the present time, over three hundred phase diagrams of silicon with other elements have been studied. Many of them have a eutectic character (for example, the systems of silicon with silver,

Table 12 Properties of silicon and germanium[37, 51, 56, 57]

Property	Silicon	Germanium
Melting point, °C	1420	936
Specific gravity, g/cm^3	2·328	5·323
Coefficient of thermal expansion at 25° C, deg.$^{-1}$*	$4·2 \times 10^{-6}$	$6·1 \times 10^{-6}$
Thermal conductivity at 25° C, cal/sec cm deg	0·20	0·14
Dielectric constant	12	16
Heat of sublimation, kcal/mole	88·04	78·44
Debye temperature (below 4° K), ° K	658 ± 6	362 ± 6
Poisson's ratio	0·28	0·21
Micro breaking strength (111), kg/mm^2	402	170
Criterion of brittleness	2·0	4·4
Forbidden gap width at 0° K (extrapolated data), eV	1·21	0·78
Electron mobility at 300° K, cm^2/V sec	1550	4400
Resistivity of the purest samples at 300° K, ohm cm	16 000	60
Ionization energy of impurity atoms, eV	~0·04	~0·01

*Of interest is the temperature dependence of the coefficient of thermal expansion of silicon, which becomes negative below 120° K[58]

aluminium, tin, gallium, indium, antimony, etc.). Silicon forms chemical compounds with lithium, phosphorus, arsenic, manganese, iron, cobalt, nickel, sulphur, selenium, magnesium, and certain other elements[61]. The silicon-copper phase diagram has also been studied in the region of very low concentrations of copper[40]. None of the mineral acids alone attack silicon, but alkalis attack it to evolve hydrogen and form salts of silicic acid. Because of its chemical inertness and tendency to form oxide films on its surface, crystalline silicon is difficult to etch.

Various etchants such as KOH, HNO_3 + HF + $AgNO_3$, etc., are used to remove layers whose structure has been damaged as a result of cutting, grinding, and mechanical polishing, as well as all sorts of surface contaminants. A method has been proposed for the electropolishing of silicon in KF – KCl solutions, as well as etching involving the use of surface active agents such as berberine[62].

Various etchants are recommended for revealing the dislocations on monocrystalline samples, such as the etchant known in the literature under the name of CP – 4 slightly modified by the addition of a small amount of mercuric nitrate see[39(p.241)] and others, for example[37(p.200), 40(p.165)].

The fusion of silicon is associated with a sharp increase in electric conductivity up to values characteristic of liquid metals. This is accompanied by an increase in density, since the structure of the short-range order is rearranged in the direction of a higher coordination number[63].

The technology of preparation of pure silicon as well as its semiconducting properties have been described in numerous studies. The collections[37, 39, 40, 62] contain basic data on the subject, the most important of which we shall cite below.

Monocrystalline silicon is used in the preparation of semiconductor devices. The raw material is pure silicon which is most often obtained by the reduction of silicon tetrachloride by zinc in the vapour phase. Zinc vapours or hydrogen can reduce other silicon halides and silane. A very efficient method is also the one involving the thermal decomposition of silicon halides on tantalum wire in the presence of hydrogen[64]. All these processes require a thorough preliminary purification of the original halide or silane by fractional distillation or chemical means.

The polycrystalline silicon thus obtained, which possesses intrinsic conduction at room temperature, is subjected to crucibleless zone melting in order to obtain single crystals of high purity. This method has substantial advantages over the others, since silicon is extremely reactive at high temperatures. In this method, the zone of molten silicon is kept stationary between two silicon rods whose ends are fixed, so that no crucible is necessary for the melt. Single crystals are thus obtained which are about 30 cm long and 1 cm in diameter, and have a high resistance ($\approx$ 3000 ohm cm). The distribution coefficients of the various impurities in silicon and germanium, which determine the degree of purification by metallurgical methods, and the diffusion coefficients of these impurities, are given in Table 13[41].

The diffusion coefficients are important in the technology of manufacture of $p-n$ junctions, which were mentioned above. The table also gives tetrahedral covalent radii, which are obviously related to the magnitude of the distribution coefficient K.

The solubility of impurities in silicon is low. The elements of the third and fifth main subgroups of the periodic system usually enter into the lattice of silicon, forming substitutional solid solutions and producing acceptor and donor levels, respectively. Other elements, such as lithium, copper, and iron, form interstitial solutions. This is what causes great differences in their diffusion coefficients (see Table 13). Neutral impurities in silicon are hydrogen, inert gases, germanium, tin and lead. Aluminium, phosphorus and boron are the most common impurities in silicon.

All the donor and acceptor impurities of elements in groups III and V can be removed with relative ease by zone melting, including

Table 13 Distribution and diffusion coefficients of impurity atoms in silicon and germanium

Element	Atomic weight	Tetrahedral covalent radius, Å	$K = C_s/C_j$ Silicon	$K = C_s/C_j$ Germanium	D cm/sec (850°C) Silicon	D cm/sec (850°C) Germanium
Si	28·09	1·17	1	3		
Ge	72·60	1·22	0·3	1		5×10^{-13}
Li	6·940			$> 10 \times 10^{-2}$	5×10^{-6}	1×10^{-5}
Cu	63·54	1·35	4×10^{-4}	$1·5 \times 10^{-5}$	4×10^{-5}	4×10^{-5}
Ag	107·88	1·53		$10^{-4} - 10^{-6}$		9×10^{-7}
Au	197·0	1·50	$2·5 \div 3 \times 10^{-5}$	$1·5 - 3 \times 10^{-5}$	1×10^{-7}	$1·5 \times 10^{-9}$
Zn	65·38	1·31		1×10^{-1}	$10^{-6} - 10^{-7}$ (1100°C)	6×10^{-12}
B	10·82	0·89	$6·8 \times 10^{-1}$	~ 20	$\sim 10^{-12}$ (1100°C)	4×10^{-12}
Al	26·98	1·26	$1·6 \times 10^{-3}$	1×10^{-1}		
Ga	69·72	1·26	4×10^{-3}	1×10^{-1}		6×10^{-13}
In	114·76	1·44	3×10^{-4}	$1·1 \times 10^{-3}$		8×10^{-13}
Tl	204·39	1·47		4×10^{-5}		8×10^{-13}
Sn	118·70	1·40	0·02	0·02		
Pb	207·21	1·46				
P	30·975	1·10	4×10^{-2}	$1·2 \times 10^{-1}$		2×10^{-11}
As	74·91	1·18	7×10^{-2}	4×10^{-2}		1×10^{-10}
Sb	121·76	1·36	$1·8 \times 10^{-2}$	3×10^{-3}	10^{-13} (1100°C)	8×10^{-11}
Bi	209·0	1·46		4×10^{-5}		
Pt	195·09		$\sim 1 \times 10^{-5}$	$\sim 1 \times 10^{-6}$		
Mn	54·94			7×10^{-3}		
Cr	52·01			2×10^{-6}	$4·8 \times 10^{-5}$	4×10^{-5}
Fe	55·85			2×10^{-6}		
Co	58·94			$1·8 \div 2·3 \times 10^{-6}$		
Ni	58·71		10^{-5}	5×10^{-6}		
Ta	180·95		1×10^{-7}			
W	183·86		10^{-5}			

phosphorus and aluminium. An exception is boron, which has a distribution coefficient close to unity (see Table 13). A method involving the zone refining of silicon in hydrogen containing water vapour has been worked out for the removal of boron[39].

It should be kept in mind that in addition to the usual impurities silicon contains a large quantity of dissolved oxygen ($1 \cdot 18 \times 10^{18}$ oxygen atoms/cm^3 of silicon), which may have an adverse effect on the electrical parameters. On the other hand, it is thought that in certain cases oxygen binds the dislocations which decrease the lifetime of the charged carriers (by forming unreactive SiO_2 molecules), and therefore has a beneficial effect[65]. Oxygen increases the lattice spacing of silicon[66]. The problem of the effect of oxygen on the electric properties of silicon should not be considered to have been finally solved.

The preparation of silicon with intrinsic conduction requires a concentration of electrically active impurities of less than 10^{10} cm^{-3}. This is a complex technical problem and is the reason why, even though silicon began to be used before germanium, development was slower because an impurity concentration in silicon of less than 10^{13} cm^{-3} has not so far been achieved.

The mobility of the charge carriers in silicon at high temperatures depends on the scattering by the lattice vibrations, and at low temperatures, by the impurity ions. Heat treatment and pressure affect the electric properties of silicon considerably.

The longwave absorption edge of silicon is at $1 \cdot 2$ μ, which corresponds to a forbidden gap width of $1 \cdot 1$ eV. The same value is obtained from experiments on the temperature dependence of the carrier concentration in the region of intrinsic conduction. Pure silicon is transparent in the range of wavelengths above $1 \cdot 5$ μ.

Silicon is widely employed as the material for semiconductor devices. Its large forbidden gap width allows these devices to operate at high temperatures. For instance, silicon rectifiers (alloy diodes) can operate at temperatures up to 200°C. It is also used in the manufacture of high-power rectifiers, amplifiers (transistors) and devices for converting the energy of radioactive fission into electrical energy[41]. The development of silicon photocells (solar batteries) of high efficiency (11%) has brought closer the solution of an important problem in present day energetics: the direct conversion of solar energy into electrical energy.

4. Silicon Carbide

Silicon carbide is a colourless, transparent substance. The β-modification of silicon carbide crystallizes in the structure of diamond (sphalerite), whilst the normal α form is hexagonal. β-SiC has a lattice spacing $a = 4 \cdot 358$ Å.

α-SiC has 7 modifications which differ in the ordering of the layers.

The number of layers in the various modifications varies from 4 to 87. In hexagonal modifications, the lattice spacing attains 1500 Å (for the silicon carbide modification with a 594-layer spacing)[48].

Specific gravity 3·217 g/cm^3
Moh's hardness of silicon carbide 9·5
Microhardness (Knoop) H_{25} 2880 ± 350 kg/mm^2

In reference 50 mention is made of the existence of covalent and ionic bonding in silicon carbide, determined from the absorption of residual radiation. The cleavage in silicon carbide is observed along (001), which does not indicate a very high proportion of ionic bond character. A survey of its properties is given in the books[67, 68]. We cite here only a few brief data.

Melting point about 2600°C (above 2200°C
SiC decomposes into the elements)
Coefficient of linear expansion (in the temperature
range 100 – 700°C) 4 – 7 × 10^{-6} deg^{-1}
(reference 45 (vol. 1 p. 220))
Refractive index[69] from 2·654 to 2·657
Dielectric constant[33] 6·7

Silicon carbide is completely insoluble in pure acids. It is attacked only by a mixture of concentrated hydrofluoric and nitric acid. When fused with alkalis, silicon carbide decomposes rather easily to form salts of silicic and carbonic acid (see reference 11 (p. 492)).

The manufacture of silicon carbide is complicated by its high melting point. The impurity content in commercial carbide obtained in electric furnaces by the reduction of silicon dioxide by carbon is usually no less than 1%. The silicon carbide obtained has different colours which vary from green to dark blue and violet, depending upon the type of impurity and on the excess of carbon or silicon over the stoichiometric ratio. Colourless crystalline silicon carbide is obtained by laboratory methods. Fine crystals of silicon carbide can be obtained on a red hot filament by reaction between benzene and silane vapours.

Silicon carbide is also obtained by the thermal decomposition of ethylene in a quartz tube at 1200 to 1300°C or by heating powdered silicon in a graphite crucible under vacuum, or else by using Verneuil's method. It can be obtained from solutions by reacting silicon tetrachloride with hydrocarbons at high temperature[67] and by other methods[41, 42, 45, 68–71].

Silicon carbide is a semiconductor with a forbidden gap width $\Delta E = 2·86$ eV (from optical measurements up to 300°K).

The carrier mobility at room temperature is
μ_n – 100 cm^2/Vsec; μ_p – 10 cm^2/Vsec.

Electronic conduction appears (together with a green colour) in the presence of impurities of group V elements: phosphorus, arsenic, antimony, and bismuth, and also iron; it is also produced by nitrogen present as the impurity.

Hole conduction appears, together with a blue colour, when elements of groups II and III – calcium, magnesium, boron, aluminium, gallium, and indium – are present as impurities. The silicon present in excess of the stoichiometric ratio causes electronic conduction, and excess carbon causes hole conduction. Recently, a region of intrinsic conduction of silicon carbide was obtained[72].

The ability of silicon carbide to withstand high temperatures makes it a very promising semiconductor. Until recently, it was used primarily in the preparation of nonlinear resistances, resistances for heating elements (frequently used as sources for infrared radiation) and lightning arresters[42].

In the last 3 to 4 years, data have appeared in the literature on the preparation of power rectifiers from silicon carbide which operate at 650°C [41].

5. Germanium

Germanium is a light grey substance with a metallic lustre. In 1871, D. I. Mendeleev predicted the existence of this element and called it ekasilicon. In 1886, Winkler discovered an element which he called germanium and which was found to be the analogue of silicon predicted by Mendeleev. After the discovery and study of the basic properties of germanium, nothing was known for many years about its potential technical applications.

During the period between 1940 and 1945, the demands of war technology led to the development of methods for the purification of germanium, and its semiconducting properties, about which little was known, were investigated. At the present time, owing to the rapid growth of technical applications of germanium and the development of research, it may be regarded as the most thoroughly studied of the semiconductors.

Germanium crystallizes in the structure of diamond; the cleavage in the crystals is the same as in those of silicon.

Lattice spacing[48]   5·658 Å*
Moh's hardness   6
Microhardness (Knoop) H_{25}

|| $(1\bar{1}0)$ 780 ± 79 kg/mm^2
|| $(11\bar{2})$ 845 ± 26 kg/mm^2

---

* Here and below the data given in X.U. ($\sim 10^{-11}$ cm) in the original literature have been converted into Å.

PHYSICOCHEMICAL & ELECTRIC PROPERTIES 73

The microhardness of germanium was studied in references 73 and 74, and the anisotropy of the microhardness was studied in reference 73. It was found that the data on the microhardness are in agreement with the characteristics of plastic deformation; the least plastic direction is (111) in the crystal and this corresponds to the greatest microhardness in the (111) plane; the most plastic direction is (110), corresponding to the smallest microhardness in the (110) plane.

Below 400°C germanium is very brittle, it becomes plastic at 600°C and at 800°C the single crystal of germanium buckles and twists[38, 40].

Reference 74 gives data on the effect of illumination on the microhardness of germanium, silicon and indium antimonide. The authors found that the illumination of the surface decreases the microhardness of the surface layer, which is 1 to 2 µ thick, by 10 to 70%. Tables 12 and 13 list certain properties of germanium.

According to the data of reference 75, the coefficient of linear expansion of germanium becomes negative below 48°K. The vapour pressure of germanium at temperatures above 370°K was studied in reference 37 (p. 161). Table 14 gives data on the vapour pressures of germanium and silicon at various temperatures. Similar data are given, for comparison, for elements which are either impurities or additives[76].

On melting, the electrical conductivity of germanium increases sharply, becoming 13 times greater. Its density also increases, as does the coordination number, which becomes equal to 8 [63].

Germanium dissolves in molten metals, zinc, aluminium, etc. It gives continuous solid solutions with silicon only. As a rule, the

Table 14 Vapour pressure of germanium, silicon and certain other elements at various temperatures

Element	Temperature in °C for a vapour pressure of (in mm Hg)							
	10^{-7}	10^{-6}	10^{-5}	10^{-4}	10^{-3}	10^{-2}	10^{-1}	1
Germanium	877	947	1037	1142	1262	1407	1582	1997
Silicon	997	1082	1177	1282	1407	1547	1713	1927
Silver	530	588	757	832	922	1032	1167	1337
Aluminium	737	807	882	972	1082	1207	1347	1547
Arsenic	127	150	174	204	237	277	317	372
Gallium	572	622	688	757	842	937	1057	1197
Indium	547	604	670	747	837	947	1077	1242
Phosphorus	62	82	107	130	157	183	222	262
Sulphur	—17	—1	18	40	66	97	125	183
Antimony	307	342	382	427	477	542	617	753
Zinc	148	172	208	242	290	342	405	485

solubility in solid germanium is low (as is the solubility in solid silicon).

To date, studies have been made of the phase diagram of binary systems of germanium with zinc, cadmium, aluminium, gallium, indium, thallium, antimony, bismuth, tin, silver, gold, and lead. These diagrams are eutectic in character; in certain cases, the eutectic is degenerate. The formation of chemical compounds has been ascertained in the reaction of germanium with sulphur, selenium, tellurium, arsenic, phosphorus, magnesium, manganese, iron, cobalt, nickel, antimony, lithium and silver. Phase diagrams of germanium-copper systems have also been studied in the region of very small concentrations of impurities, of the order of 10^{16} cm^{-3} [40].

Germanium fails to react with carbon even at high temperatures, and this makes it possible to use graphite boats for the treatment of germanium; nor does it react with quartz, which is an advantage in the technology of its preparation.

Germanium is stable at room temperature, but rapid oxidation begins above 600°C. An oxide layer about 1000 Å thick is formed fairly rapidly on thin polycrystalline films of germanium in air, whilst the layer is only 20 Å thick on single crystals under the same conditions. The kinetics of oxidation at room temperature are reported on in reference 60. Germanium dissolves in a mixture of hydrofluoric and nitric acid, in alkali with hydrogen peroxide, and also in aqua regia. In the electromotive series it lies between copper and silver.

The dissolution of the surface of germanium and the nature of the associated chemical processes have been studied very thoroughly[78-82]. Study of the kinetics of dissolution in nitric acid has suggested a possible mechanism for the oxidation of germanium by molecules of dinitrogen tetroxide with hydrated molecules of $GeO(OH)_2$ going into solution. Statistically, the most definite results were obtained from a study of the dissolution of germanium in alkaline solutions of hydrogen peroxide and solutions of bromine and iodine[78,79].

Various compositions are recommended for etchants of germanium; electropolishing in a solution of potassium fluoride or sodium chloride and nitric acid and other methods are used for cleaning the surface[40,62]. The etchant CP-4 is employed most frequently for revealing dislocations[37 (p.188)]. There are also selective etchants, for example, the etchant "WAg" [38 (p.177)]; they form etch figures bounded by planes which are parallel to certain well-defined crystallographic planes.

In order to obtain germanium with intrinsic properties at room temperature, it is necessary to purify it until the concentration of impurities is less than 10^{13} cm^{-3}. The solution of this problem was a major step forward in the technology of semiconductors. The technology of preparation of pure monocrystalline germanium and of germanium doped with various admixtures has been developed in detail[38,39,40]. The principal stages are as follows:

The initial germanium is dissolved in hydrochloric acid to separate the impurities. The germanium tetrachloride obtained is distilled in the presence of chlorine to remove traces of arsenic and is then hydrolysed. The germanium dioxide which precipitates is washed, dried, and reduced with hydrogen between 650 and 675°C. The germanium powder obtained is fused and subjected to zone crystallization (these processes are often carried out in the same apparatus). A given quantity of the necessary impurity is then added to the purified germanium, the mixture is fused in an inert atmosphere, and the monocrystalline specimen is withdrawn from the melt with the use of a seed. The orientation of the crystal is determined by the seed.

At the present time, the method of "zone levelling" has found wide applications in the preparation of the starting material for the production of many types of transistors and diodes. In zone levelling, a molten zone containing the necessary doping impurity is passed along a polycrystalline ingot. A single crystal seed is placed at the origin of the zone, so that the substance solidifies as a single crystal.

In order to obtain a germanium single crystal with a very uniform distribution of impurities along its length, the melt is replenished during the pulling.

A technique has now been developed which makes it possible to evaluate the degree of perfection of the structure of a germanium single crystal. A correlation has been established between the electric properties and the density of the dislocations in the crystal. The density of the dislocations may be determined from the density of the etch pits which are formed where the dislocations intersect the surface of the crystal. Experimentally, it is possible to establish a certain permissible maximum, and to select a material by making use of this limit. Obviously, coarser irregularities (grain boundaries, twins, etc.) should be excluded.

Above we have discussed the mechanism of the effect of acceptor impurities of group III and of donor impurities of group V in germanium and silicon. Many chemical impurities, e.g., copper and nickel, give rise in germanium to additional, very effective centres which shorten the lifetime of the charged carriers. Literature data on the effect of oxygen in germanium are contradictory.

Elements of groups III and V dissolve quite well in germanium and have small diffusion coefficients (see Table 13). As far as the elements of other groups which act as carrier recombination and trapping centres are concerned, they are soluble in germanium to a considerably lesser extent and have large diffusion coefficients. Neutral impurities, i.e. those which do not cause the formation of new carriers in germanium, include hydrogen, inert gases, nitrogen, silicon, tin, and lead[40].

Since the technology of purification of semiconducting materials and semiconductor manufacture is related to diffusion processes, special literature has been devoted to the subject[83]. More recently investigations have been conducted into the process of electro-diffusion in semiconductors, which is interesting in that it is related to all the electric and diffusion properties of substances. The electro-diffusion of antimony and indium in germanium is discussed in reference 84.

In germanium single crystals of the highest purity, the carrier lifetime is apparently limited not only by the structural defects and dislocations, but by the residual content of impurities like copper and nickel as well. As compared to silicon, however, germanium is generally much easier to purify by zone levelling, since it has no impurities with a distribution coefficient close to unity (see Table 13).

The problem of the influence of lattice defects has lately assumed great importance. Thus, the problem of the interaction of dislocations and impurities of copper and other elements is being studied in great detail[39 (p.85)]. Because of the development of tunnel diodes, which require large impurity concentrations in the materials, the problem of the behaviour of impurities in concentration ranges far removed from the impurity levels usually considered in semiconductors has become urgent. This aspect of the problem has linked the fields involving the research on micro- and macro- impurities in semiconductors.

The electric properties of pure germanium are given in Table 12. In the range of high temperatures, the carriers in germanium are scattered predominantly by the lattice vibrations, whereas scattering by impurities is prevalent at low temperatures.

The heat treatment of germanium has a considerable effect on its electrical properties. Abundant literature[38-40] has been devoted to this problem. Pressure also alters the electrical parameters considerably.

Like silicon, germanium is transparent in the infrared. At wavelengths less than 1·8 µ, germanium absorbs, particularly in the visible portion of the spectrum. A comparison of the curve for the spectral distribution of the photoconductivity of germanium with the curve for the optical absorption shows that the photoconductivity maximum coincides with the wavelengths at which the absorption coefficient begins to drop off sharply. This region corresponds to the activation energy of the intrinsic carriers, i.e. to the forbidden gap width.

Germanium has found wide applications in semiconductor electronics. It is used to manufacture diodes, transistors, transducers, photoresistances, optical filters, α-particle counters, power rectifiers and other devices.

6. The Silicon – Germanium Solid Solution

Silicon and germanium form a continuous series of substitutional solid solutions corresponding to Roozeboom's type I phase diagram, as was established[86]. Subsequently, a thermodynamic calculation of the silicon – germanium phase diagram was carried out, and its details were refined[87]. Fig. 10 shows a phase diagram of these alloys, based on experimental and calculated data.

The main characteristic of this system of solid solutions is the irregularity of the crystallization process, the cause of which is to be found in the characteristics of the covalent type of the chemical bond in silicon and germanium, particularly its directivity and rigidity which hinder the interdiffusion of silicon and germanium atoms. The process of nonequilibrium crystallization in analogous cases was discussed in reference 90. The authors of reference 86 obtained a homogeneous solution with difficulty; they accomplished this by heating for 5 to 7 months at a temperature close to that of the liquidus and by regrinding and repressing the alloys. Later, homogenerous alloys of this solid solution were obtained by zone melting[62 (p.59)] and other methods[88 (p.89)].

In experiments using the method of zone melting, conditions were created for equilibrium crystallization by using a slow rate of zone travel. In order to obtain a region of solid solution of the same composition and of considerable length, and to increase the size of the grains of the alloy, the molten zone was passed through the specimen several times, and the direction of motion was varied. Later a method was proposed for preparing crystals of the solid solution Si – Ge from the gas phase[91].

The microhardness of the alloys of the solid solution of silicon and germanium under a 100 g load was studied by means of the Knoop indenter[37 (p.427)]. The hardness values were considerably scattered, but it was possible to conclude that there was no maximum on the hardness versus composition curves: the variation is close to linear. This fact is explained[92] by the proportionality of the hardness to the yield point and the relation of the latter to the shear modulus.

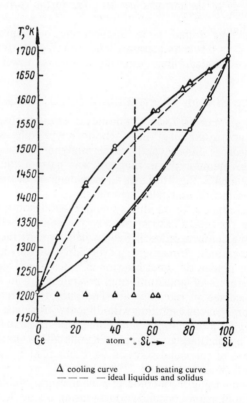

Fig. 10 *Phase diagram of the Ge–Si solid solution*[87]

In reference 37 (p. 414), the lattice constants and the forbidden gap were determined by the optical method in homogeneous alloys subjected to polarographic analysis in order to define the composition more accurately. Figs. 11 and 12 show the variations of the lattice constant and forbidden gap width with the silicon concentration.

The variations in density showed that the latter, as well as the lattice spacing, changes linearly with the composition within the limits of accuracy. As can be seen from Fig. 12, the forbidden gap width changes linearly with the composition up to 10 atom per cent of silicon. This is followed by an appreciable change in the slope of the straight line. Hence, small amounts of silicon added to ger-

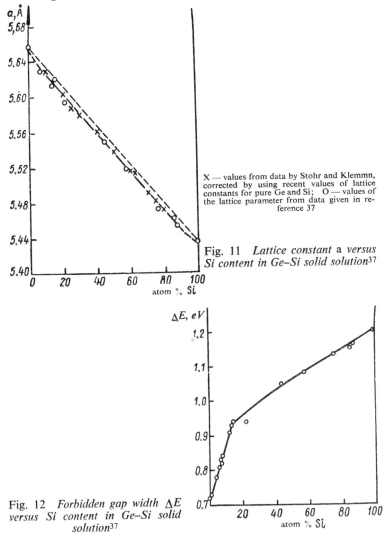

X — values from data by Stohr and Klemmn, corrected by using recent values of lattice constants for pure Ge and Si; O — values of the lattice parameter from data given in reference 37

Fig. 11 *Lattice constant a versus Si content in Ge–Si solid solution*[37]

Fig. 12 *Forbidden gap width ΔE versus Si content in Ge–Si solid solution*[37]

manium increase the forbidden gap width considerably, and this is of practical importance[93]. An attempt at a theoretical interpretation of this course of the change in the forbidden gap width with the composition is given in reference 94.

Studies of the magnetic and other properties of silicon and germanium have revealed that their energy spectra are different. Herman[94] holds that the addition of silicon to germanium causes a change in the relative energy levels of the valence band and the conduction band. This is what determines the non-linear course of the variation in the forbidden gap width in solid solutions of silicon in germanium. The similarity in the energy spectra of the initial substances should produce a linear variation of the forbidden gap width during the formation of solid solutions.

The career mobility was studied in silicon-germanium alloys[57,95]. Table 15 lists the mobilities of electrons as functions of the alloy composition.

The silicon-germanium alloys, the properties change continuously, so that when the required material is chosen, it is possible to combine the advantages of each of the initial components, to produce, for example, a wider range of operating temperatures than in germanium, and a higher mobility than in silicon.

Table 15 Carrier concentration and electron mobility in silicon-germanium alloys

Silicon content of alloy, %	Carrier concentration, cm^{-3}	Electron mobility, $cm^2/V\ sec$
0	10^{15}	4400
6—7	1.5×10^{15}	2330
14	1.9×10^{15}	525
25	2.9×10^{16}	370
98	9.7×10^{15}	1150
100	10^{15}	1550

7. Grey Tin

Grey tin (α-modification) is a loose, dark grey powder. It crystallizes in the structure of diamond.

White tin (β-modification), a high-temperature modification of tin, crystallizes in the tetragonal structure and is a metal, whereas grey tin is a semiconductor.

Table 16 lists the properties of the two modifications of tin for comparison[96-98].

Table 16 Properties of the α and β modifications of tin

Property	α-Sn	β-Sn
Lattice spacing, Å	$a = 6.4912 \pm 0.0005$	$a = 5.8197$ $c = 3.1749$
Specific gravity, g/cm³	5.765	7.298
Coefficient of linear expansion, deg⁻¹	20.9×10^{-6}	5.0×10^{-6} (from -130 to $+25°$ C)
Forbidden gap width, eV	0.08	—
Debye temperature, °K	230	—

As seen from Table 16, the densities of the two modifications differ rather strongly (of all the elements, only carbon shows such a great difference in the densities of its modifications). Therefore, the conversion of white tin into grey tin is accompanied by an appreciable change in volume. This makes it difficult to obtain grey tin in the form of single crystals. The coefficient of linear expansion of grey tin at 45°K becomes equal to zero and passes into the region of negative values.

In absolute value at room temperature, the coefficient of linear expansion is almost four times smaller than that of white tin; this, as well as the density difference, is due to the special features of the crystal structure of grey tin, and of the considerably covalent character of the bond between its atoms. From the standpoint of kinetic relationships, the probability of metastable states is associated with the stresses arising in the course of the formation of the new phase, which are in turn determined by the difference in the specific volumes of the initial and final phases[99]. This interpretation provides a satisfactory explanation of the metastability of white tin.

As was shown by studies of the magnetic properties of the two modifications, the paramagnetic white tin becomes diamagnetic when converted into the α-modification. This is also consistent with representation of grey tin as a substance continuing the series diamond – silicon – germanium.

The temperature of transition of white tin into grey tin is $+13.2°C$. When 0.75% germanium is added to grey tin, the latter may be stable for days, even at $+60°C$[100]. The linear rate of conversion of white tin into grey tin at $-20°C$ is 0.003 mm/hr, the maximum rate being at $-30°C$[101]. The rate of conversion is found to depend on the medium, the impurities, the heat treatment of the tin, the mechanical strains, the supercooling, and certain other factors.

Grey tin dissolves in dilute hydrochloric, sulphuric, and nitric acid, as well as in hot alkaline solutions. The first studies of the electric properties were initiated by A. F. Ioffe[102], who suggested that it has the properties of a semiconductor, since, from the standpoint of its structure and chemical bonding between its atoms, it should belong to the series of semiconducting substances diamond – silicon – germanium. Ioffe's suggestion was confirmed. Grey tin was found to be a substance whose properties completed the series of diamond-like semiconductors of subgroup IVb of the periodic system.

Further investigations of the electrical properties were accompanied by investigations aimed at improving the method of its preparation[103-105]. Filaments[106-107] and single crystals were obtained from a liquid solution in mercury. In the latter work, the authors obtained grey tin with hole or electron conduction by varying the impurities in the mercury. According to references 97, 102-107, the electric properties of tin show that it is a semiconductor with a narrow forbidden gap width ($\Delta E = 0.08$ eV) and a high carrier mobility $\mu_n \approx 300$ cm^2/Vsec. The electron mobility at 100°K attains 30,000 cm^2/Vsec.

A study was also made of the photoconductivity of grey tin at low temperatures. The limit of photoconductivity was found to be at about 12 µ, which corresponds to a forbidden gap width of the same order as the one indicated above.

Because of its poor stability grey tin has not found practical application. However, its investigation has been of great scientific importance, since it confirmed the relationship between the electrical properties of semiconductors and their chemical nature, i.e., the link between position in the periodic system, crystal structure, bond type and semiconducting properties. It made it possible to make predictions about the electrical properties of substances still unknown. This was undoubtedly very important for the further development of both the chemistry and the physics of semiconductors.

An interesting and practically significant result obtained from investigations of grey tin was the discovery that it was possible to "contaminate" white tin not only with grey tin, as had been known earlier, but also with cadmium telluride and indium antimonide – substances which at that time were completely unknown as semiconductors.

On the basis of these observations, which indicated not only a structural similarity but a similarity in bond type between grey tin and the above-mentioned binary compounds, a hypothesis was advanced regarding the semiconducting properties of indium antimonide and cadmium telluride which was confirmed experimentally[96,108].

These studies were the beginning of extensive investigations into the group of binary diamond-like semiconductors, performed since the early fifties and up to the present in many laboratories, both in the Soviet Union and abroad.

Part B

Binary Compounds

1. Compounds of Type A^3B^5

Binary compounds of type AB belonging to the crystallochemical group of diamond-like semiconductors are related by their physicochemical nature. The properties typical of the entire group are manifested most clearly by compounds of type A^3B^5, formed by atoms whose positions in the periodic system are closest to group IV. Conversely, the compounds A^1B^7 formed by atoms which are most distant from group IV resemble salts in certain properties (for example, in colour). Whilst the study of compounds A^3B^5 is relatively recent, the compounds A^2B^6 and A^1B^7 were studied by chemists a long time ago.

All the substances A^3B^5, A^2B^6, and A^1B^7 are poorly soluble in water; this usually indicates a nonpolar (covalent or metallic) bond. Most of the compounds A^3B^5 and A^2B^6 are stable to oxygen and atmospheric moisture (except the compounds of aluminium and beryllium), and they are all derivatives of hydrides and liberate the latter upon decomposing.

The crystal structure of all diamond-like compounds is being studied at the present time, and their space groups are being determined with great accuracy. For some of them, the existence of second modifications has been established (see Table 3); sometimes the second modifications exist only in thin layers.

Cleavage has been studied in most of the compounds which are of interest. In addition to the standard goniometric methods, the "razor" technique of investigating microhardness has been used for semiconducting substances of the diamond group[50].

The hardness of the compounds in this group was first studied by Goldschmidt, who used Moh's scale; the microhardness of many diamond-like substances has since been studied.

In an overwhelming majority of cases, binary diamond-like compounds are the only compounds which form in the given systems with a component ratio of 1:1. This appears to be a natural consequence of the fact that it is precisely in binary compounds formed by elements equidistant from the middle of the periodic system that the normal valence of atoms is satisfied and that a symmetrical structure of the valence shells of the atoms arises which is similar to the configuration of the shells of inert gases. The average number of electrons per atom becomes four, which ensures the possibility of crystallization in a structure similar to that of the elements—diamond and its analogues. As we shall see below, this latter similarity may be followed in many properties of binary compounds.

As can be seen from Figs. 13 to 19, the phase diagrams of the compounds A^3B^5 are of the same type. The equilibrium gas phase of these compounds consists mainly of the more volatile component. Despite an appreciable dissociation, in almost all of the compounds of type A^3B^5 studied from this point of view, a strong tendency to form compounds of stoichiometric composition is observed upon solidification. This is apparent from the fact that it is possible to obtain indium antimonide, indium phosphide and gallium arsenide with a concentration of impurity atoms which is so low that it could not be observed as any appreciable deviation from stoichiometry (keeping in mind that superstoichiometric atoms and foreign impurities are equivalent as far as the formation of electrons is concerned). Gallium antimonide is an exception in this respect; it is possible that its chemical composition does not correspond to the stoichiometric one.

The solution of the problem involving the width of the region of homogeneity in compounds of type A^3B^5 undoubtedly requires the use of many methods. Thus far, precision X-ray analysis of indium phosphide, indium arsenide and gallium arsenide has provided no indication of the presence of an appreciable region of homogeneity in the vicinity of the stoichiometric composition[109,110]. Nor could any deviation from stoichiometry be detected in indium antimonide by studying its electrical properties[111-113].

Because of this special feature, in preparing compounds of type A^3B^5 in the monocrystalline state and purifying them, it has been possible to use the same methods which were used for a similar purpose in the case of germanium and silicon, i.e., withdrawing from the melt, zone melting, etc.

The behaviour of some substances of this group in the molten state is interesting. For instance, the fusion of the antimonides is accompanied by an increase in density, which corresponds to a rearrangement of the short-range order structure in the direction of increase in the co-ordination number and approach to close packing. This is also a manifestation of the similarity in the properties of elemental semiconductors and compounds of type A^3B^5.

Compounds of type A^2B^6, insofar as can be judged from the available experimental data, behave differently in the molten state, showing a tendency to form molecular structures. No data of this kind are available for 1^1B^7.

Compounds of type A^2B^6 as well as A^3B^5 can be obtained in the form of thin films by sublimation under vacuum, and, as has been shown by electron diffraction studies, some of these substances form crystals of the hexagonal modification of the wurtzite type in addition to the usual crystals with a zinc blende structure.

For most substances of type A^3B^5, the kinetics and mechanism of the action of various reagents have been studied very little thus far. In etching, suitable etchants must be selected empirically. Use is often made of the etchant developed for germanium (CP-4) in various concentrations.

Many characteristics of the compounds A^3B^5 have also been studied relatively little despite their importance. These include thermal conductivity, heat of formation, etc.

It should be noted that most of the physicochemical investigations of diamond-like substances were initiated by the requirements imposed by practical applications. For example, studies of the $P - T - x$ diagrams of A^3B^5 compounds with volatile components were carried out because of the necessity of subjecting these compounds to zone recrystallization, in order to obtain them in an ultrapure state. It is conceivable, therefore, that in the near future, because of the expanding applications of diamond-like substances, their physicochemical characteristics will be studied as exhaustively as their electrical properties. The latter have been treated in a very large number of studies, which is growing continuously[114,115]. There is no doubt that this great interest is elicited not only by scientific but also by purely practical considerations.

In every field of technological application, there arises a series of specific requirements with respect to the basic characteristics of a semiconducting material. Such characteristics, as was mentioned above, are the forbidden gap width, the concentration and degree of ionization of the impurities, the mobility, the carrier lifetime, etc.

From the standpoint of these basic parameters, semiconductors of type A^3B^5 are of great interest as materials for rectifiers, amplifiers, photocells, Hall e.m.f. transducers, etc. Certain compounds of type A^2B^6 also find applications in radio electronics. However, the properties of substances in these groups might also be used in other areas of application where substances unrelated to the group of diamond-like semiconductors have been employed until now. Thus, for example, data have appeared on the feasibility of using certain compounds of type A^3B^5 in thermoelectric generators[116]. Especially promising in this area are multicomponent substitutional solid solutions in which it is possible to obtain a small thermal conductivity coupled with a high electric conductivity and thermoelectromotive force[117].

In studying the effect of impurities on compounds of type A^3B^5, a definite similarity has been observed with the behaviour of impurities in germanium and silicon.

As a rule, in binary semiconductors A^3B^5, the elements of group II, zinc and cadmium, function as acceptors, i.e., give rise to hole conduction, while the elements of group VI, sulphur, selenium, and tellurium, function as donors, giving rise to electronic conduction. There are also elements whose mechanism of action is due to the penetration of their atoms into the interstices, and also electrically neutral impurities. However, all these problems (compared with the same problem for silicon and germanium) have been studied to a considerably smaller extent for binary diamond-like semiconductors, although the number of such investigations is growing rapidly.

Until comparatively recently, germanium was the only semiconductor obtained in such a pure form that it could be used as the basic material for the quantitative check of the theory, and for making numerous electronic semiconductor devices. At present, such an "ideal" semiconductor is ultrapure indium antimonide.

In addition to new physicochemical properties, as compared to the properties of diamond, silicon, germanium and grey tin, compounds of type A^3B^5 also possess a new combination of basic electrical parameters which are characteristic only of this group of substances and are most clearly expressed in its individual representatives.

A very characteristic feature of these compounds is a high electron mobility, a relatively low hole mobility, and a small effective mass of charge carriers. All these properties enable one to observe, in some of them, certain new, still unexplained phenomena: anomalies in optical absorption, a weak temperature dependence of the Hall effect and of the electrical conductivity in the range of low temperatures, a negative change in resistance in a magnetic field, and others.

The discovery and investigation of semiconductors possessing a very high electron mobility made possible the actual use of the Hall effect in such devices as semiconductor magnetic field strength gauges. Indium antimonide, indium arsenide, mercury selenide and mercury telluride were used for such purposes.

At the present time, the most interesting and practically significant prospects of application, in addition to those mentioned above, are those of indium antimonide and indium arsenide as materials for infrared detectors, and those of indium phosphide as the material for rectifiers and amplifiers. Gallium arsenide appears to be more promising as a material for solar batteries than silicon. Gallium arsenide and aluminun antimonide will undoubtedly find broad applications as materials for rectifiers[113].

Among binary semiconductors which have been studied little or not at all, of the greatest interest from the standpoint of practical applications is probably gallium phosphide as a material for rectifiers operating at high temperatures, and boron compounds which should have the same kind of advantages over gallium arsenide as silicon carbide exhibited in comparison with germanium[118].

A more extensive development of semiconductor devices based on binary compounds is hindered by the fact that the methods of synthesis and purification of these compounds have not yet reached the high level of, for example, those used for germanium. In general, the history of the introduction of germanium into technology repeats itself for its analogues in a somewhat different version. Just as in the case of germanium, the promise of its application in semiconductor technology is determined primarily by the nature of the substance, but the actual feasibility of the application is determined by the purity and perfection of the crystals.

In the case of comparatively low-melting diamond-like compounds, for instance indium antimonide and gallium antimonide, the synthesis does not involve any difficulties; they are obtained by one of the most popular methods, the combined fusion of the components in evacuated and sealed quartz ampoules in a regular furnace[96]. The most complex stages are the purification of the starting materials, and the further purification of the compounds and their preparation in the monocrystalline state. A very delicate task is the doping of the semiconductor materials and the concomitant problems of homogenization.

The problem of obtaining the starting materials in a state of high purity is solved in the most diverse ways at the present time: use is made of the processes of precipitation, extraction of the impurities with solvents, vacuum fractional distillation, thermal dissociation, etc.[119,120].

For the metallic components of the semiconducting compounds of this group, as well as for the semiconducting materials of type A^3B^5 themselves, use is usually made of the method of fractional recrystallization proposed in 1949 by Likhtman and Maslennikov for the preparation of single crystals[121], and developed further for germanium, resulting in variants existing at the present time and employed in purification[122, 123].

The method most commonly used for the purification of monocrystalline samples is that due to Czochralski. As is indicated above, all these processes are based on the use of the difference in the compositions of the adjacent solid and liquid phase of the substance during crystallization, and on the forcing of the impurities away from the crystallization front. The purification achieved by withdrawing the single crystal from the melt is more effective than the purification achieved in one passage in zone recrystallization. However, the latter may be made more productive, so that in practice substances are subjected to repeated zone recrystallization, and single crystals are then pulled out of the pure melt.

The methods used for the purification and preparation of single crystals of binary semiconductors are generally analogous to those used in the technology of preparation of ultrahigh-purity germanium. However, owing to the volatility of the non-metallic component of the arsenides and phosphides, the methods of their preparation and purification have been altered somewhat in order to eliminate the adverse influence of the evaporation of arsenic and phosphorous.

The process of evaporating the volatile component may go so far as to lead to a substantial decomposition of compounds into their constituents. Other methods of synthesis are therefore available which amount to an independent regulation of the vapour pressure of the volatile component.

The so called "two-temperature" method[124] is often employed for the synthesis of complex arsenide and phosphide semiconductors. It includes the following basic steps:

1. The metallic and the non-metallic component are located in different parts of the ampoule, and the heating is done separately.

2. The metallic component is heated to a temperature which is slightly higher than the melting point of the compound.

3. The non-metallic component, taken in excess, is heated to the temperature at which the necessary vapour pressure of this component is reached.

Very frequently, the processes of synthesis and zone recrystallization are combined and the method of crucibleless zone recrystallization in a sealed vessel has been successfully applied to decomposing A^3B^5 compound[125]. In introducing a definite amount of impurity and adjusting the composition, use is made of a method which involves the displacement of the molten zone through a solid ingot, sometimes in two opposite directions alternately. This method has proved very fruitful when applied to complex semiconductors.

Among the diamond-like semiconductors there exists, however, a group of very refractory substances to which the method of the combined fusion of the components taken in stoichiometric amounts is completely inapplicable, or else gives unsatisfactory results. Furthermore, modern technology requires particularly refractory semiconducting materials whose properties remain unchanged up to very high temperatures. This has led to the development of methods of preparation of semiconducting materials from the gas phase (of particular interest is the method of transport reactions in which a halogen is used as a transporting agent for the semiconductor atoms) and from flux-melts.

The latter method, first applied to semiconducting substances by Wolff, Keck and Bröder[126] consists of the following: an evacuated quartz ampoule is filled with 80 to 90% of the non-volatile metal component (Al, Ga, In), which acts as a flux, and 10 to 20% of the non-metal (P, As, Sb), and the ampoule is sealed. The heating is carried to temperatures slightly higher than the liquidus temperature. During subsequent slow cooling, the compound precipitates in the form of crystals which are separated from the excess metal by dissolving the latter. The crystallization process usually produces crystals that are purer than the initial substances.

A substantial lowering of the synthesis temperature is achieved in this method, as compared with the melting point of the compound; this follows from the phase diagram. Since the temperature is lowered, the probability of contamination by the impurities contained in the material of the reaction vessel is decreased. A major advantage of the method is the decrease of the pressure in the ampoule, and this method can therefore be used with substances whose volatile component has a high vapour pressure.

In many cases, the methods used in preparing semiconductors from the gas phase may have certain advantages (for instance, if the substances must be obtained in the form of thin films or needles). Since compounds of type A^3B^5 are stable in most cases, then, despite the fact that their composition includes such poisonous substances as arsenic, their use in semiconductor devices under normal conditions is completely safe. However, when these materials are handled in synthesis, in mechanical or thermal treatment, or when their properties are studied, it is necessary to keep in mind that the products of reaction of these materials with atmospheric moisture, which are poisonous even in minute amounts, are easily absorbed through the skin, and that the dustlike particles may enter the respiratory tract. It is therefore necessary to observe a number of safety rules, to use rubber gloves and plexiglas chambers, etc.

A description of the physicochemical properties of all the compounds of type A^3B^5 known to date is given below. Also given are the basic parameters which characterize the semiconducting properties of these substances, i.e. the forbidden gap width, the carrier mobility, and certain others.

Boron Compounds

Boron nitride, BN, is known in two modifications.

One modification, which is graphitelike in structure, is called "white gel" and is a white flaky powder whose individual crystals are transparent.

The crystal strucure of this hexagonal boron nitride differs from the graphite structure in that the graphitelike layers are arranged exactly underneath one another.

Space group[48] $C6m2$ (D_3^{12}), $z = 2$
Lattice spacing[127] $a = 2\cdot504$ Å
$c = 6\cdot661$
Specific gravity [128] $2\cdot25$ g/cm^3

The electron density distribution is similar to that in graphite, but the distance between the lattices in boron nitride is less than that in graphite (BN — $3\cdot34$ Å, graphite — $3\cdot40$ Å). This leads to a different structure of the energy spectrum and to the disappearance of metallic properties.

Owing to the characteristics of the crystal structure and the low hardness (about 2 units on Moh's scale) boron nitride possesses lubricating properties[129].

Several methods of preparation of hexagonal BN are known. A very pure product can be obtained by the method due to Podszus, improved by Samsonov et al.[130, 131]. The method consists of treating with ammonia, at 1000 to 1200°C, a B_2O_3 charge based on the sinter $B_2O_3 + CaO$, obtained by heating boric acid with chalk. Boron nitride obtained by roasting borax with ammonium chloride is used for the preparation of luminophors[132].

Boron nitride melts at 3000° (under nitrogen pressure) and is stable in a neutral or reducing atmosphere up to this temperature. It hydrolyses under the influence of moist air or dilute acids.

Dissociation vapour pressure 9 mm Hg at 1220°C,
158 mm at 2045°C, and 760 mm at 2500°C[129]
Characteristic temperature 598°K
Heat of formation 60·7 kcal/mole
Forbidden gap width 4·6 – 3·6 eV

At 300°C, the thermal conductivity is $0\cdot036 - 0\cdot069$ cal/cm sec deg, and at 1000°, $0\cdot029 - 0\cdot064$ cal/cm sec deg. The coefficient of thermal expansion in the range of 25 to 350°C is $10\cdot15 \times 10^{-6}$, and in the range of 25 to 1000°, $7·5 \times 10^{-6}$. These last two parameters drop as the temperature rises, which distinguishes BN from metals. The dielectric constant at a frequency of 10^2 Mc/s and a temperature of 10°C is 4·15. The resistivity, $1·7 \times 10^{13}$ ohm cm at 25°C[133], decreases with an increase in temperature.

In the infrared spectrum, there are two absorption bands at wavelengths of 7·28 and 12·3 μ, which correspond to the two principal crystallographic directions in the BN crystal[134].

Detailed data on the methods of preparation and the mechanical and thermal properties of BN are given in monograph[129], and data on BN considered as a luminophor are given in monograph [135]. BN is used as a thermal insulator and a refractory lubricating compound. The potential use of its semiconducting properties is promising.

Diamond-like BN, named borazone by the American scientists who obtained it[128], is almost as hard as diamond. Borazone crystals have colours ranging from yellow to black or are colourless; its crystal structure belongs to the sphalerite type (zinc blende).

Space group $F\bar{4}3m$, $z = 4$
Lattice spacing (at 25°C) $3·615 \pm 0·001$ Å
Specific gravity $d_{exp} = 3·45$ g/cm^3
$d_{x-ray} = 3·47$ g/cm^3
Energy of B – N bond 35 kcal/mole

The presence of another compound, $B(N_3)_3$[136], has been discovered in the B – N system. This compound, boron triazide, is a white explosive powder. There is also evidence indicating the existence of B_3N and $B(N_3)_3$[137].

Borazone is obtained by heating hexagonal boron nitride at 1360°C and 62,000 atm pressure [128], and also by nitrogenating boron phosphide[138].

Acids do not attack borazone. When heated to 2700° under vacuum, it does not undergo any changes and becomes slightly oxidized at 2000° in air, in contrast to diamond, which burns as low as 900°.

From the standpoint of its electrical properties, borazone is a dielectric, and its uses in technology will apparently be broader than those of diamond, owing to the great heat stability of the nitride, the hardness being almost the same.

Boron phosphide is transparent in the crystalline state, is reddish in colour, and is yellowish-brown in the form of a cake or powder. It is known only in one modification, which has the structure of sphalerite.

Lattice spacing 4·537 Å

Length of the B – P bond 1·964 Å
Specific gravity[139] d_{exp} 2·89 g/cm^3
$d_{x-ray} = 2·97$ g/cm^3
Knoop microhardness[141] H^{100} 3200 kg/mm^2

Boron phosphide has high abrasive properties and cuts glass[140].

In the system B – P[142,143], another phosphide is known, B_5P_3, which forms from B – P at 1000° in a hydrogen atmosphere. It is quite inert compared to amorphous B – P.

Reference 144 reports on a phosphide of the composition B_6P. The phosphide B_5P_3 could not be detected. Reference 145 disputes the existence of all boron phosphides except B – P and $B_{13}P_2$. Thus the data on all boron phosphides except B – P are contradictory. The equilibrium gas phase consists of the non-metallic component, and its vapour pressure at the melting point is apparently high.

The old works of Besson and Moissan[142,143] describe indirect methods of preparing BP in the amorphous or microcrystalline state. In 1957, boron phosphide was obtained by a direct reaction between boron and phosphorus in an evacuated and sealed quartz ampoule at 1100°C. The two-temperature method of synthesis was used in order to limit the vapour pressure of phosphorus[140]. Boron phosphide was obtained in similar fashion[139,146]. Vickery[138] obtained it in the form of films by the thermal decomposition of coordination compounds of the type $BCl_3 \cdot PCl_5$.

Williams and Ruherwein[144] obtained BP by four methods. The first was similar to the method used in reference 140. The second consisted of a reaction between aluminium phosphide and boron chloride at 175° C for 3·5 hr. The third method was based on the formation of a boron phosphide film by passing a mixture of BCl_3 and PH_3 through a quartz tube heated to 1000° C. The same films were obtained in the fourth method, in which phosphine was replaced by hydrogen and phosphorus. Stone and Hill[114], using vapour-phase reactions and crystallization from the melt, obtained needle-shaped crystals 1 to 2 mm long and granules 1 mm in diameter.

The chemical properties of amorphous boron phosphide and those of the crystalline material differ considerably. The latter is characterized by a greater chemical inertness, and does not dissolve in acid or in boiling aqueous alkaline solutions. It can be dissolved only in melts of alkalis. None of the known etchants attacks it. It is stable in air up to 800 – 1000° C.

Heat of formation[147]   49 kcal/mole
Forbidden gap width [141,144] 5·9 eV

Index of refraction 3·0 – 3·5
Thermo-emf 300 mV/deg

The (maximum) absorption coefficient for a noncrystalline sample is 10^3 cm, which occurs at 12·1 μ. The rectification factor (at a point contact) is of the order of 1000 and is retained up to 400°C [141].

Boron phosphide appears to be very promising as a material for high-temperature rectifiers.

Boron arsenide is a dark-grey substance crystallizing in the structure of sphalerite. The lattice spacing is 4·777 Å. Cubic arsenide is distinguished by its great hardness and its abrasive properties. It is stable upon heating in arsenic vapours and is stable to corrosion. The equilibrium phase consists of arsenic, whose pressure at the melting point should be high.

Boron arsenide is obtained as a powder or cake by heating arsenic with boron in evacuated and sealed ampoules at 800° for 12 to 50 hr[144,146]. In the latter work there are indications of the formation of lower arsenides.

Boron antimonide is not known as yet[147], although there are no apparent reasons preventing its formation. It is possible that the process of synthesis of BSb is made difficult by its extreme inertness. It should be a semiconductor of great hardness, should have a large forbidden gap width and a melting point of about 2000°C.

Aluminium Compounds

Aluminium nitride is a white or light-grey powder crystallizing in the structure of wurtzite.

Lattice spacings[148]: $a = 3·104 \pm 0·005$ Å
$c = 4·965 \pm 0·008$ Å
$c/a = 1·600$

Specific gravity 3·26 g/cm^3
Melting point (at 4 atm pressure in nitrogen) 2200°C
Heat of formation[51] 57·7 kcal/mole

Aluminium nitride may be obtained by the action of nitrogen on aluminium powder at temperatures above 800° C, and by heating aluminium powder in ammonia. When heated in air, AlN begins to oxidize at approximately 1200°C, and when heated in vacuum, dissociates at 1750°C[149]. In water, the nitride slowly decomposes, evolving ammonia. Aluminium nitride is a luminophor and a semiconductor. The forbidden gap width is 3·8 eV according to the data of reference 150, while other authors hold that it should be greater than 5 eV.

In view of the fact that it has not yet been obtained in the form of large crystals, the area of its application in semiconductor technology has not yet been defined.

Aluminium phosphide is a yellowish-grey mass. It crystallizes in the structure of sphalerite.

Lattice spacing[151]   5·46 Å
Moh's hardness   5·5
Forbidden gap width[153]   3·0 ± 0·3 eV

The phase diagram of the aluminium – phosphorus system has not been studied. Data are available according to which the phosphide is insoluble in aluminium. The melting point should be about 2000°C, and the equilibrium vapour pressure at the melting point should be high.

The phosphides Al_3P_7, Al_5P_3, Al_3P, have been described, but their existence has not been definitely established.

The aluminium phosphide AlP was first obtained by Goldschmidt, who passed phosphorus vapours in a stream of hydrogen over aluminium powder at 500°C[151]. It was subsequently obtained by roasting a mixture of aluminium powder and red phosphorus powder[152]. It has been obtained, by Addaniano[125], by melting aluminium with zinc phosphide Zn_3P_2 in a sealed ampoule at 800 to 900°. Recently, AlP was obtained, in an apparatus made of corundum, in the form of yellow needles 1 mm long and 0·2 mm thick[155].

The phosphide is readily decomposed by water, acids, and alkalis with the evolution of phosphine, and does not melt or dissociate when heated to 1000°C. According to the data of reference 155, $\Delta E AlP = 2.42$ eV at 20°C, and rectification, photo-emf and electroluminescence are observed.

Aluminium arsenide is a dark substance with a metallic lustre. As a powder it has a reddish brown tinge. It crystallizes in the structure of sphalerite.

Lattice spacing[154]   5·6622 Å
Moh's hardness   5
Microhardness[118] H_{50} 500 ± 20 kg/mm^2
Characteristic temperature[154]   400° K
Melting point   > 1600°C
Coefficient of linear expansion[154]   3·5 × 10^{-6} deg^{-1}
Forbidden gap width[153]   2·16 ± 0·1 eV

The phase diagram of the system Al – As is given in Fig. 13[156]. Aluminium arsenide does not dissolve the components in appreciable amounts. The eutectic with arsenic is completely degenerate, as is that with aluminium.

The formation of AlAs takes place in an atmosphere of arsenic vapour, whose pressure has not been determined. Formerly, AlAs was obtained by melting the powdered starting materials in evacuated quartz ampoules and grinding and heating once again. Lately, the two-temperature method[157] has been used for the synthesis of this compound. Most convenient is the method involving the melting of the components (not powdered) with vibrational stirring[158].

Aluminium arsenide is decomposed by water and by moist air; its electrical properties have scarcely been studied at all.

Aluminium antimonide is a dark-grey substance with a metallic lustre. It crystallizes in the structure of sphalerite.

Lattice spacing[159] 6.1355 ± 0.0001 Å
Microhardness H_{50} 400 kg/mm^2 [118]
Knoop microhardness H_{25} 395 ± 34 kg/mm^2 [49]
Specific gravity (exp) 4·15 g/cm^3
Heat capacity $C_p = 2.8$ cal/g atom deg
Characteristic temperature[160] 350° K
Heat of formation[166] 3 ± 2.5 k cal/mole
Index of refraction[167] 3·0

According to the data of [49], aluminium antimonide has a cleavage plane not only along (001), but also an additional imperfect cleavage along (111) and between (111) and (011), which in the author's view indicates a smaller percentage of the ionic component and a larger percentage of the covalent component.

The phase diagram of aluminium antimonide was first explained in detail by G. G. Urazov[161], and it indicated the presence of only one compound, AlSb, with a melting point of 1060°C. Urazov came to the conclusion that the reaction between antimony and aluminium proceeds to completion if the alloy is kept, while thoroughly stirred, at a temperature 100 to 150°C above the temperature at which the crystallization begins. The author explains the difficulty of reaching an equilibrium in this reaction by the chemical inertness of antimony.

Fig. 14 shows the phase diagram of the system Al – Sb. The author established the insolubility of the components in the compound with an accuracy of 0·1% by weight. Later, Guertler and Bergman[162] found that a low solubility does exist, but that it decreases with a decrease in temperature. They also established the presence of 2 eutectics, on the side of aluminium at 650°C and on the side of antimony at 624°C. The same authors determined the melting point of aluminium antimonide to be 1050° C.

AlSb expands on crystallizing, and the volume increase is approximately 13%[163]. V. M. Glazov studied the physicochemical properties of aluminium antimonde in the liquid state during the precrystallization period, and also when the substance was somewhat superheated. On the basis of a study of the viscosity, electrical conductivity and magnetic susceptibility, Glazov reached the conclusion that the reaction involved in the formation of aluminium antimonide is a second-order one.

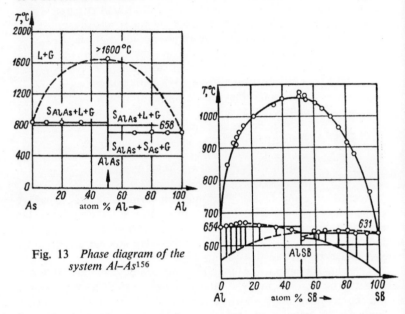

Fig. 13 *Phase diagram of the system Al–As*[156]

Fig. 14 *Phase diagram of the system Al–Sb*[161]

The compound is stable at the melting point and at this temperature the vapour pressure of antimony does not exceed 10^{-3} mm Hg[164]. During the melting and within a small range above the melting point, the strong chemical interaction between the components is preserved. During the precrystallization period, the coordination number, density and metallic character of the bond are decreased[165].

Aluminium antimonide differs from gallium antimonide and indium antimonide in its appreciable reactivity, and it is therefore more difficult to prepare. Quartz ampoules cannot be used for this synthesis, since they are quickly destroyed. Graphite crucibles are more convenient, but aluminium antimonide reacts with graphite, although not so vigorously, forming aluminium carbide, Al_4C_3. It is more desirable, therefore, to use crucibles and boats made of corundum (sintered Al_2O_3).

Aluminium antimonide is obtained by melting stoichiometric amounts of aluminium and antimony in evacuated ampoules[168] or in an argon atmosphere[169]. Zone recrystallization in an atmosphere of pure argon is an efficient method for purifying it and this is used to produce the starting material for the preparation of pure single crystals by withdrawal from the melt (also in an argon atmosphere[1])[69, 170].

The impurities Ge, Cu, Ag, Au, Li, Be and Sb do not change the conduction type of aluminium antimonide. Zn and Cd are acceptors, while Sn, Pb, As, Bi, Te, Se are donors[168, 170 etc].

After zone melting and the withdrawal of single crystals, aluminium antimonide usually possesses hole conduction. In order to obtain electronic conduction, tellurium or selenium are added[168]. In practice zone recrystallization does not improve the resistance of the samples; this is due to the fact that in aluminium antimonide the impurity levels are produced by a "disorder in the lattice"[171] or by vacancies in the lattice of antimony[72], and not by foreign contaminants. Apparently, this also explains the fact that the degree of purity of aluminium antimonide achieved thus far is low.

The electrical properties of aluminium antimonide are characterized by the following figures:

Forbidden gap width of $0°K$ (from optical data) $1\cdot6$ eV
Hole mobility   400 cm^2/V sec (300° K)
   < 5000 cm^2/V sec (50° K)
Electron mobility[172, 173]   200 cm^2/V sec

Aluminium antimonide is subject to considerable corrosion in moist air, but it has been observed that the higher the purity of the material, the better it withstands corrosion. On the other hand, the introduction of indium antimonide or gallium antimonide and the formation of a solid solution slows down the corrosion substantially, and prevents it when the concentration of the mixture is high (see Part C, Section 1). The mechanism of corrosion is described by the overall equation[124]:

$$AlSb + 3H_2O \rightarrow Al(OH)_3 + Sb + 3/2 H_2.$$

In dry air, no corrosion occurs. Despite the fact that, in comparison with gallium antimonide and indium antimonide, aluminium antimonide possesses undesirable properties (reacts with quartz and graphite and is subject to corrosion), the technology of aluminium antimonide is developing, since the forbidden gap width of AlSb lies in the region of values characterizing the materials which are promising for use in solar batteries. AlSb devices exhibit a rectifying effect with a rectification factor of $\approx 10^4$. For this reason, aluminium antimonide, from which diodes are now being made, has wider prospects of application.

The preparation of aluminium antimonide in thin films is possible. However, the properties of such films have not been satisfactory thus far. For this reason, layers of AlSb do not have practical applications[164, 174].

GALLIUM COMPOUNDS

Gallium nitride is a fine powder, which, depending on the method of preparation, has a light grey or yellowish colour. Gallium nitride crystallizes in the structure of wurtzite.

Lattice spacing $a = 3 \cdot 180 \pm 0 \cdot 004$ Å
$c = 5 \cdot 160 \pm 0 \cdot 005$ Å
Specific gravity[175] $d_{exp} = 6 \cdot 10$ g/cm^3
$d_{x-ray} = 6 \cdot 11$ g/cm^2
Melting point $\sim 1500°$ C
Heat of formation $24 \cdot 9$ kcal/mole
Forbidden gap width[150] $3 \cdot 25$ eV

It is obtained as a loose grey powder by heating gallium metal in a stream of ammonia at 1200°C[176], and as a yellowish powder by the decomposition of $(NH_4)GaF_6$[177]. Upon heating in air, it begins to oxidize at 1000°, and when heated in vacuum, it dissociates at 1050°C[149]. GaN dissolves poorly in acids and alkalis[152].

Gallium phosphide has the form of transparent orange crystals with a metallic lustre. It is known in only one crystalline form which has the structure of sphalerite.

Lattice spacing[159] $4 \cdot 4505 \pm 0 \cdot 0001$ Å
Moh's hardness   5
Knoop microhardness H_{25} 945 ± 155 kg/mm^2
Microhardness (sq. pyram) H_{50} 940 ± 35 kg/mm^2
Coefficient of linear expansion $3 \cdot 5 \times 10^{-6}$ deg^{-1}
Index of refraction[172] $2 \cdot 9$

Gallium phosphide has a cleavage plane only along (011), which indicates a considerable ionic character of the compound as compared with its analogues[49].

The phase diagram of the gallium – phosphorus system has been studied insufficiently, but is thought to be similar to the phase diagram of the indium – phosphorus system (see Fig. 17). The melting point is considered to be approximately 1500°C at a pressure of ~ 25 atm[180]. As do most compounds consisting of a non-volatile metallic component and a volatile non-metallic component, gallium phosphide decomposes upon melting into an alloy rich in the non-volatile component and a vapour rich in the volatile one.

Gallium phosphide is obtained by melting the elements in an

evacuated and sealed quartz ampoule with vibrational stirring[158]. If the alloy is crystallized slowly at one end, it is possible to obtain a compact, transparent, orange substance with a metallic lustre[157]. Monocrystalline needles up to 12 mm long and 1·5 mm thick are obtained by crystallization from excess gallium[126].

Recently, indirect methods of preparing gallium phosphide have been proposed: (1) reaction of gallium monochloride or monoiodide with yellow phosphorus or phosphine[178], (2) reaction between gallium and zinc phosphide, the zinc being subsequently driven off[125], and (3) reduction of gallium trioxide by phosphine[179].

In the last two cases, gallium phosphide is obtained as a yellow powder, and in the first case in the form of small crystals. Upon heating in air, gallium phosphide begins to oxidize at about 750°C, and when heated under vacuum, it dissociates at approximately 1000°C[164]. On the basis of these data, high-pressure equipment has been developed for the preparation of crystalline ingots[180].

Gallium phosphide is a semiconductor. Obtained by melting, it usually possesses hole conduction, and electronic conduction can be obtained by adding sulphur. It displays a good rectification at a point contact, and photoconductivity[114].

Forbidden gap width at 0° K (from optical measurements)[172] 2·4 eV
Electron mobility[182] > 110 cm^2/V sec
Hole mobility[182] > 75 cm^2/V sec

It combines interesting semiconducting and luminescent properties. When a d.c. or a.c. current is passed through single crystals of gallium phosphide, electroluminescence is observed; depending on the activator, the light emitted has an orange or a dark-red colour[181]. It is promising as a material for high-temperature rectifiers; its use in the optical and electron- and magneto-optical areas is also possible[164, 182].

Gallium arsenide is a dark grey substance with a violet hue and a tarry lustre. It crystallizes in the structure of sphalerite.

Lattice spacing[159] 5·6534 ± 0·0002 Å
Moh's hardness 4·5
Knoop microhardness[49] H_{25} 750 ± 42 kg/mm^2
Microhardness (sq. pyram) H_{50} 700 kg/mm^2
Specific gravity (X-ray) 5·4 g/cm^3
Shear modulus [56] 3·6 × 10^5 kg/mm^2
Young's modulus [56] 9·1 × 10^5 kg/mm^2
Poisson's ratio [56] 0·29
Micro breaking strength 189 kg/mm^2
Criterion of brittleness 3·0

Index of refraction[172] 3·2
Characteristic temperature 303°K
Thermal conductivity (at 300°)[365] 125 × 10^{-3} cal/cm sec deg
Dielectric constant[154, 183] 11·1
Coefficient of linear expansion[172] 5·7 × 10^{-6} deg^{-1}

In addition to the cleavage along (011), GaAs has a well-defined cleavage along (111) and between (111) and (011), which shows that it has the lowest proportion of ionic bonding in any of the compounds of type A^3B^5 [49].

It was shown in reference 184 that as the temperature decreases the coefficient of linear expansion α decreases, becomes zero at $T \sim 55°K$, and then shifts to the region of negative values. The smallest value of α is observed at $T \sim 40°K$ and is equal to -0.5×10^{-6} deg^{-1}; at 300°K, $\alpha = 5\cdot8 \times 10^{-6}$ deg^{-1}.

The phase diagram of the system gallium – arsenic is given in Fig. 15[185], which shows the $T-x$, $P-T$ and $P-x$ projections of the system gallium – arsenic, where T is the temperature in °C, x is the concentration in atomic %, and P is the pressure in atm. As is apparent from Fig. 15, gallium arsenide has no phase transitions; its melting point is 1237 ± 3°C. The eutectics on the sides of the components are degenerate. At the melting point, the pressure of arsenic is ~ 0·9 atm.

Table 17 gives values corresponding to a three-phase $P-T-x$ diagram, where T is the temperature at which solid gallium arsenide is at equilibrium with the melt and the vapour, and T_1 is the temperature (see Fig. 15) corresponding to the pressure of arsenic vapour P_{As}[186].

The width of the region of homogeneity of gallium arsenide cannot be determined by the X-ray method[110].

The compound is obtained by melting the components in an evacuated and sealed quartz ampoule. The use of the two-temperature method of synthesis with vibrational stirring has great advantages, since it makes possible the synthesis of large amounts of the substance (of the order of hundreds of grams) in the course of 1 to 1½ hr[187]. Zone melting is often carried out together with the synthesis by the two-temperature method. The elements Cd and Zn act as acceptors in gallium arsenide, and S, Se, and Te act as donors[188].

The withdrawal of single crystals from the melt by Czochralski's method, with seeding, is sometimes modified so that, instead of the stoichiometric melt, use is made of a melt with a high content of the non-volatile component, and the temperature of the withdrawal of the non-volatile component is lowered.

Gallium arsenide can also be obtained by the free growth of

Table 17 Pressure, temperature and concentration in phase diagram 'solid phase-liquid-vapour' of the gallium-arsenic system

T_1, °C	P_{As}, atm	T, °C	Arsenic content of melt (in atomic %)
386	6.2×10^{-3}	781 ± 20	7.5
438	1.8×10^{-2}	895 ± 20	10.5
485	5.2×10^{-2}	1068 ± 1	19.0
492	6.05×10^{-2}	1055 ± 3	18.0
508	8.9×10^{-2}	1085 ± 5	20.5
532	1.55×10^{-1}	1181 ± 3	31.0
543	2.01×10^{-2}	1190 ± 3	33.0
562	3.2×10^{-1}	1196 ± 3	34.5
569	3.8×10^{-1}	1221 ± 3	38.0
600	7.6×10^{-1}	1234 ± 3	46.0
616	1.18	1235 ± 4	55.0
645	1.95	1231 ± 4	57.5
673	3.35	1205 ± 5	64.5
711	6.60	1185 ± 5	68.5
810*	2.9×10	810	100.0

*Degenerate eutectic point. The partial pressure of arsenic at this point is equal to the pressure of pure arsenic at the given temperature.

crystals in a melt containing excess gallium[126]. Chemical methods are employed less often. Using such a method, which involved passing hydrogen saturated with arsenic vapours over gallium trioxide, Goldschmidt obtained gallium arsenide for the first time[151]. The method of preparation from the gas phase, mentioned above in the description of gallium phosphide, has been proposed for the preparation of this substance[178].

The etching of gallium arsenide for the purpose of revealing the microstructure is done with 3 per cent hydrogen peroxide and ammonia in the proportion of 1:3. A special etchant[189] has been developed for cleaning the surface prior to the deposition of contacts, and also to make $n - p$ junctions.

Under ordinary conditions, gallium arsenide is stable toward atmospheric moisture and oxygen. Upon heating in air, it begins to oxidize at 600° C, and begins to dissociate in vacuum at about 850°C[164]. Gallium arsenide is a semiconductor.

Forbidden gap width at 0°K 1.53 eV
Electron mobility[182,191] 400 – 8500 cm²/V sec
Hole mobility[182,191] 200 – 400 cm²/V sec

The forbidden gap width of GaAs is closer to the wavelength of the maximum of solar radiation than is the forbidden gap width of silicon. For this reason, gallium arsenide is used as a material for solar batteries, in which solar energy is directly converted into electrical energy by means of the photoelectromotive force[190].

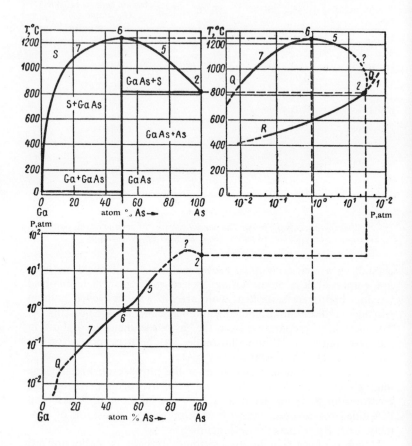

Fig. 15 *T–x, T–P and P–x projections of the system Ga–As*[185]

Gallium arsenide displays good rectification on point contacts[114]. The electrical conductivity and Hall constant are practically totally independent of temperature in the range of 1·5 to 300°K for electronic samples of gallium arsenide. This is due to the fact that the substance is in a degenerate state (i.e. metallic) at an electron concentration of $n = 10^{17} - 10^{18}$ cm^{-3} [192].

It is used to make junction-type rectifiers, tunnel diodes and triodes, and detectors for visible and X-ray radiation[193-196]. Because of the large forbidden gap width, devices made of the arsenide operate at higher temperatures than those made of germanium and silicon. Gallium arsenide is a semiconductor with a great technological future. Tunnel diodes based on it are particularly promising for radio engineering applications.

Gallium antimonide is a light grey alloy with a metallic lustre. It crystallizes in the structure of sphalerite.

Lattice spacing[159]   6·0954 ± 0·0001 Å
Moh's hardness 4·5
Knoop microhardness 448 ± 27 kg/mm^2
Microhardness (sq. pyram) 420 ± 10 kg/mm^2
Specific gravity (X-ray) 5·65 g/cm^3
Shear modulus 2·90 × 10^5 kg/mm^2
Young's modulus 7·60 × 10^5 kg/mm^2
Poisson's ratio 0·30
Micro breaking strength 151 kg/mm^2
Criterion of brittleness[56]   1·8
Heat capacity (at 80°K) 3·5 cal/g atom deg
Heat of formation[198]   4·97 kcal/g atom
Debye temperature (at 80°K)[160] 270
Melting point[197]   712 ± 0·3°C
Index of refraction[172]   3·7
Linear coefficient of expansion 6·9 × 10^{-6} deg^{-1}

In addition to the cleavage plane along (011) there is an additional imperfect cleavage along (111) and between (111) and (011), which indicates a smaller contribution of the ionic component and a large contribution of the covalent component to the bonding[49].

A phase diagram of the gallium-antimony system, taken from reference 156, is shown in Fig. 16.

It follows from this diagram that the system contains only one compound, GaSb, which has a melting point of 703°C. The eutectic of the antimonide with antimony (13 atom % gallium) melts at 583°C, and the eutectic of the antimonide with gallium is degenerate. The vapour pressure of antimony at the melting point of gallium antimonide is approximately 10^{-2} mm Hg.

The width of the homogeneity region of gallium antimonide cannot be determined even when a precision X-ray technique is used[110].

Fusion of the material is accompanied by an increase in density which corresponds to a rearrangement of the short-order structure in the direction of an increased co-ordination number. The melting process is also accompanied by a sharp increase in electrical conductivity, up to values corresponding to the conductivity of pure metals[63].

The volume change taking place when gallium antimonide melts is equal to 7 per cent. At the melting point, it dissociates to a negligible extent[199]; under ordinary conditions, it is stable toward the moisture and oxygen of air, and oxidation begins at approximately 400°C [164].

The etching of the surface for the purpose of revealing the microstructure is done with concentrated nitric acid or a 5 per cent solution of ferric chloride in dilute (1:2) hydrochloric acid[200].

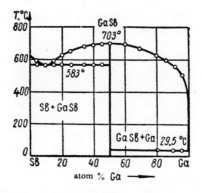

Fig. 16 *Phase diagram of the system Ga–Sb*[156]

Gallium antimonide is obtained in the form of monocrystalline needles from excess gallium[126]. The method most commonly employed involves the melting of the components in an evacuated and sealed quartz ampoule. The synthesis is sometimes carried out in a hydrogen atmosphere.

Coarse-grained ingots are obtained directly from the synthesis when slow cooling is used. Single crystals are readily obtained by Czochralski's method of withdrawal and also by other methods (Bridgman and others).

Zone recrystallization yields a material which possesses hole conduction and does not lend itself to a high degree of purification. Material having electronic conduction (as well as $n-p$ junctions) can be obtained by doping hole-type gallium antimonide with selenium or tellurium. The material purified by zone melting gives 10^{17} acceptors per cm^3 at best, regardless of the preliminary purification of the components. The hypothesis was therefore advanced that the impurity which is not eliminated during the zone melting is not a foreign element, but gallium at the sites of antimony. This is equivalent to admitting the presence of a compound of which the composition is substantially different from stoichiometric[201]. However, no direct proof of this hypothesis has thus far been obtained.

Forbidden gap width 0·80 eV
Electron mobility[172]   4000 cm^2/V sec
Hole mobility[172]   700 cm^2/V sec

The semiconducting properties of GaSb are close to the properties of germanium. However, since gallium antimonide has not been obtained in a state of high purity thus far, it is of no particular interest from the standpoint of practical applications. It is possible that it will find applications as a material for tunnel diodes. As will be shown below, gallium antimonide may prove to be useful as a constituent of complex semiconducting alloys. Thus, for instance, its addition to aluminium antimonide prevents the latter from corroding, and its addition to indium antimonide increases the forbidden gap width without appreciably reducing the mobility.

INDIUM COMPOUNDS

Indium nitride is a black, soft power. It crystallizes in the structure of wurtzite.

Lattice spacings: $a = 3·53 \pm 0·004$ Å, $c = 5·69 \pm 0·004$ Å
Specific gravity[175] $d_{exp} = 6·88$ g/cm^3, $d_{x-ray} = 6·91$ g/cm^3
Melting point $\sim 1200°$C
Heat of formation[177] 4·6 kcal/mole
Resistivity[202] $4·0 \times 10^{-3}$ ohm cm

Indium nitride is obtained by the decomposition of $(NH_4)_3$ InF$_6$ at 600°C. In air, it begins to oxidize at 600°C, and in a vacuum it begins to dissociate at 620° [149].

The electrical properties have been studied: the metallic character of the conductivity and a resistivity of $4·0 \times 10^{-3}$ ohm cm have been established[202]. However, these results are subject to doubt, since an approximate calculation of the forbidden gap width yields the very appreciable value of 2·4 eV [103].

Indium phosphide is a dark grey substance with a tarry lustre. It crystallizes in the structure of sphalerite.

Lattice spacing[159]   5·86875 ± 0·0001 Å
Knoop microhardness H_{15} [49]   535 ± 47 kg/mm^2
Microhardness (sq. pyram.) H_{50} [204]   435 ± 20 kg/mm^2
Index of refraction[172]   3·0
Melting point 1070°C

Indium phosphide displays cleavage along (011)[50].

Fig. 17 shows the $T-x$, $P-T$ and $P-x$ projections of the solid – liquid – vapour phase diagram of the indium – phosphorus system[185]. Since indium phosphide is a substance with a very appreciable vapour pressure of the volatile component, only a part of the In – InP diagram has been studied, as is evident from Fig. 17: the InP – P part could not be studied. The eutectic on the side of indium is completely degenerate. At the melting point, the equilibrium vapour pressure of phosphorus attains ~ 60 atm.

Table 18 gives numerical data for the parameters of the three-phase P – T – x diagram, where T_1 is the temperature of the vapour pressure of pure phosphorus, P_p is the vapour pressure of phosphorus[186], and T is the temperature at which solid indium phosphide is in equilibrium with the melt and the vapour.

The width of the homogeneity region of indium phosphide was studied by means of X-rays. It was found that its lattice spacing did not change upon the introduction of an excess of any of the components into the system. Aqua regia is used to reveal the structural features in metallographic studies[200].

Indium phosphide begins to oxidize in air upon heating at about 500°C. Dissociation in vacuum begins at approximately 750°C [164]. It can be obtained by passing phosphine in a stream of nitrogen through indium monoiodide. Indium phosphide powder is formed upon the thermal decomposition of the complex which separates out[178]. The phosphide is also obtained by heating indium metal in a stream of hydrogen saturated with phosphorus vapours[205]. Needle-shaped single crystals can be obtained from excess indium[126].

At the present time, indium phosphide is most often obtained from a melt containing a few per cent more indium than called for by the stoichiometric ratio. According to the phase diagram, the pressure of phosphorus above the melt drops to values which are safe for the usual strength of quartz apparatus.

Zone melting is carried out in the same manner by forcing the excess metal into the end of the ingot. Single crystals of indium phosphide have also been obtained by horizontal zone melting, which usually produces crystals with electronic conduction.

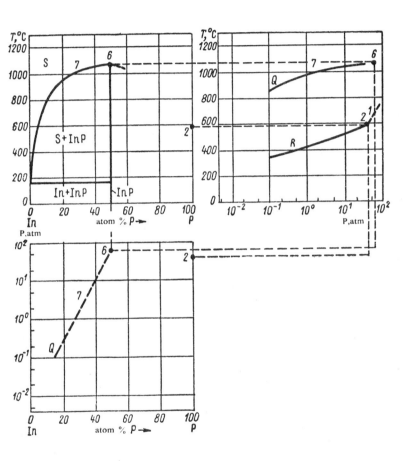

Fig. 17 T–x, T–P and P–x projections of the system In–P[185]

The distribution coefficients of the donors, sulphur and selenium, are, respectively, 0·8 and 0·6. It is thought that the distribution

Table 18 Pressure, temperature and concentration of phosphorus in the phase diagram of the indium-phosphorus system

T_1, °C	P_p, atm	$T (\pm 7)$, °C	Phosphorus content of melt atomic %
338	0·10	860	15·6
375	0·32	940	21·0
400	0·60	962	25·3
420	1·10	983	34·0
450	2·30	1013	36·5
480	4·80	1023	—
515	10·50	1030	45·9
545	20·00	1043	43·6
585	40·00	1057	47·7

coefficients of the acceptors, cadmium and zinc, are considerably smaller[124,164].

A very interesting method of preparing $n-p$ junctions in indium phosphide was proposed by Folberth[164]. It involves local heating of the melt with partial volatilization of phosphorus and a redistribution of the impurities in the solidifying phosphide according to the distribution coefficient of the impurities for another composition of the melt and another temperature. The excess indium, which solidifies last, forms the electrode.

Indium phosphide is one of the most promising semiconductors from the standpoint of practical application. Its forbidden gap width is 1·34 eV at 0°K. Until relatively recently it was obtained with a carrier concentration of $n \approx 10^{17}$ cm^3. The mobilities for this purity are $\mu_n = 3400$ cm^2/V sec at 290° K, $\mu_p = 50$ cm^2/V sec[172]. Recently, considerably purer samples have been obtained with a carrier concentration $n = 6 \times 10^{15}$ and an electron mobility of 5000 cm^2/V sec[206].

Indium phosphide displays an appreciable rectification effect and can also be used as a material for amplifiers. Its optical properties are very interesting: the absorption is less than in the purest samples of germanium.

The electron-hole junctions obtained with this material may be used as neutron detectors[114].

Indium arsenide is a dark grey substance with a metallic lustre. It crystallizes in the structure of sphalerite.

Lattice spacing[159] 6·0584 ± 0·0001 Å
Microhardness[50] 330 ± 10 kg/mm^2
Knoop microhardness H_{25} 384 ± 26 kg/mm^2

Indium arsenide displays a definite cleavage along (111) and between (111) and (011), in addition to the cleavage along (011), which indicates the smallest contribution of the ionic component of the bond in compounds of type A^3B^5 [49].

According to the data of reference 56, the properties of indium arsenide are as follows:

Shear modulus $8·20 \times 10^5$ kg/mm^2
Young's modulus $8·11 \times 10^5$ kg/mm^2
Poisson's ratio 0·24
Criterion of brittleness 2·5
Coefficient of linear expansion[172] $5·3 \times 10^{-6}$ deg^{-1}
Characteristic temperature[154] 242°
Index of refraction[112] 3·2
Heat of formation[198] 7·40 kcal/g atom

Table 19 shows the values of the pressure, equilibrium temperature, and composition in the "solid – liquid – vapour" phase diagram of the In – As system, the designations being similar to those of Table 17.

The phase diagram of the system indium – arsenic[185] is shown in Fig. 18. As is evident from Fig. 18, indium arsenide does not have any phase transitions up to the melting point, which is 942°C. The eutectic is degenerate on the side of indium, and melts at 731°C on the side of arsenic. At the melting point, the pressure of arsenic is equal to 0.3 atm.

The first investigators of the system indium – arsenic noted that indium arsenide has a very narrow region of homogeneity. The width of this region could not be accurately ascertained by X-ray analysis since an accuracy of several ten-thousandths of an angstrom is required [207].

Indium arsenide is prepared by melting the initial components in evacuated and sealed quartz ampoules which are subjected to vibration in the course of the synthesis[158]; zone refining is usually rather ineffective.

Given below are the distribution coefficients of various elements in InAs[208].

Mg ....0·7   Sn ....0·09
Zn ....0·77   S ....1·0
Cd ....0·13   Se.........0·93
Si ....0·4   Te ....0·44
Ge ....0·07

Table 19 Pressure, temperature and concentration of arsenic in the phase diagram in the indium-arsenic system

T_1 °C	P, atm	T, °C	Arsenic content of melt atomic %
438	1.53×10^{-2}	790 ± 5	20.5
473	4.10×10^{-2}	875 ± 5	32.0
520	1.18×10^{-1}	928 ± 3	44.0
532	1.71×10^{-1}	936 ± 4	48.0
553	2.45×10^{-1}	939 ± 3	49.0
576	4.8×10^{-1}	942 ± 3	51.5
600	8.3×10^{-1}	939 ± 3	57.5
619	1.02	931 ± 3	60.0
661	2.50	896 ± 3	69.0
692	4.80	868 ± 4	75.5
720	7.70	825 ± 4	81.5
735	9.70	781 ± 3	86.0
731	9.30	731	87.5

It is apparent that the distribution coefficient of sulphur is equal to unity and is very close to it for selenium; thus zone refining cannot be effective for these impurities. It is assumed that indium arsenide normally contains admixtures of these elements, chiefly sulphur, and for this reason zone recrystallization was not able to produce indium arsenide with an impurity concentration of less than 10^{17} cm^{-3}.

Apparently, in order to obtain indium arsenide with a low carrier concentration by means of zone recrystallization, in purifying the initial components it is necessary to remove the elements having distribution coefficients which are undesirable for the process of zone melting. Quite recently, Effer obtained very pure indium arsenide in this fashion[209].

Arsenic was prepared either by thermal decomposition of arsine of high purity or by reduction of 99.999% pure arsenic trioxide by pure hydrogen. Indium was purified by the usual methods. The sulphur content of the alloys after the synthesis was less than 0.00005%. The indium arsenide was then subjected to triple zone recrystallization.

Samples obtained by fusion usually possess electronic conduction. As a rule, the elements of group II introduced into indium arsenide give rise to hole conduction, and elements of groups VI and IV give rise to electronic conduction[208].

Single crystals of the arsenide are also prepared by zone melting and by withdrawing with a seed.

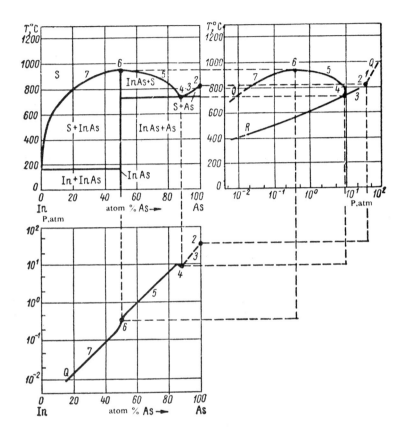

Fig. 18 *T–x, T–P and P–x projections of the system In–As*[185]

The compound is stable under ordinary conditions; when heated in air, it begins to oxidize at about 450°C, and when heated in vacuum it begins to dissociate at about 720°C.

A mass-spectrometric study of the vaporization processes has shown that the molecules As_4 and As_2 exist in the vapour phase.

The activation energy of vaporization lies between 78 and 88 kcal/mole [210].

In order to reveal structural features, indium arsenide is etched by the etchant commonly used for germanium (CP-4), which consists of five parts concentrated nitric acid, three parts hydrofluoric acid, three parts acetic acid and a few drops of bromine. Dilute boiling hydrochloric acid can also be used[164,200].

Thin layers can be obtained from indium arsenide by separate vaporization. The quality of the layers obtained depends on the temperature of the substrate, which should be approximately 700°C, in order to make the semiconducting properties of the layers correspond approximately to the properties of bulk material [211, 212]. Indium arsenide is a semiconductor.

Forbidden gap width at 0°K 0·45 eV
Electron mobility[172] 23,000 cm^2/V sec
Hole mobility[172] 240 cm^2/V sec

As was mentioned above, it was recently obtained with $\mu_n = 75 \times 700$ cm^2/V sec (at an electron concentration $n = 8 \times 10^{15}$ cm^{-3} at 77°K[109].

Experiments with the Ettingshausen-Nernst effect support the conclusion that the ionic component of the bond in indium arsenide is small (a conclusion also reached on the basis of experiments involving the determination of the cleavage); the signs of the Ettingshausen-Nernst constants were never positive as would be the case if the carriers were scattered by polar vibrations of the lattice[213,214].

The Hall constant and the electrical conductivity in indium arsenide samples with electron concentrations $n \approx 10^{16}$ cm^{-3} remains almost unchanged when the temperature changes over a wide interval (from the temperature of liquid air up to + 50°C), and this characterizes indium arsenide as a semiconductor with a degenerate electron gas (for the given conditions). These properties make it a very valuable material for the preparation of Hall generators for measuring magnetic field intensities, large direct currents, etc.[215,216]. In addition, indium arsenide is interesting as a material for infrared detectors[114], and also thermoelectric generators[116].

Indium antimonide is a light grey alloy with a metallic lustre. It crystallizes in the structure of sphalerite.

Lattice spacing[159] 6·47877 ± 0·00005 Å
Knoop microhardness[25] 223 ± 20 kg/mm^2
Microhardness (sq. pyram) H_{50} 220 ± 10 kg/mm^2
Specific gravity (X-ray) 5·78 g/cm^3
Shear modulus 2·61 × 10^5 kg/mm^2
Young's modulus 6·61 × 10^5 kg/mm^2

Poisson's ratio 0·27
Criterion of brittleness[56] 2·2
Coefficient of linear expansion $5·5 \times 10^{-6}$ deg^{-1}
Index of refraction[172] 4·1
Heat capacity $C_p = 4·08$ kcal/g atom
Characteristic temperature[160] 228° K

As in indium arsenide, the cleavage is displayed not only along (011) but also along (111) and between (111) and (011), which indicates a minimum contribution of ionic component to the bonding in these compounds[49].

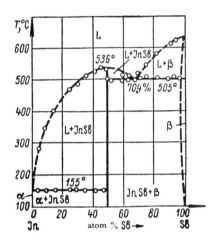

Fig. 19 *Phase diagram of the system In–Sb*[219]

It has been shown in reference 98 that the coefficient of linear thermal expansion shifts into the region of negative values at $T \simeq 5·75°$K. At 25°K, α attains the maximum absolute value, equal to $0·162 \times 10^{-5}$ deg^{-1}. Reference 217 gives the coefficient of linear expansion $α = 4·8 \times 10^{-6}$ deg^{-1}.

The heat of formation of indium antimonide, determined by fusion in a bomb calorimeter, is equal to 3·89 kcal/g atom[218].

The phase diagram of the system indium – antimony as given in reference 219 is shown in Fig. 19. As is evident from this figure, this system contains only one compound with a melting point of 536°C.

The eutectic of indium antimonide with antimony contains 70·4 atom % antimony, and crystallizes at 505°C; the eutectic with indium is degenerate. The region of homogeneity of the compound is very small, as was confirmed in the studies[111–113]. The saturated vapour pressure of solid indium antimonide was studied by the effusion method[220].

The vapour pressure above the antimonide at the melting point is approximately 10 mm Hg[164], which agrees with the data of reference in order of magnitude.

Like gallium antimonide, indium antimonide in the solid state is a substance with a predominantly covalent bond type, and in the liquid state is a metal with a predominantly metallic bond type. The fusion is accompanied by an increase in density and a sharp rise in the electrical conductivity[63]. The volume change taking place upon fusion amounts to 11·4 – 13·7%[164].

Under ordinary conditions, indium antimonide is stable toward the moisture and oxygen in air.

The surface properties were reported in references 221 – 225; it was found that different crystallographic planes may possess different properties[224].

Studies have been published on the kinetics of solution of indium antimonide in nitric acid[226] and in hydrochloric acid solutions of ferric chloride and iodine[227]. The experiments showed that, in contrast to germanium[228], in all of the investigated cases of solution of indium antimonide the process determining the rate of solution is the process of diffusion. It is obvious, therefore, that in indium antimonide the potential barrier of the process of the heterogeneous reaction is lower than that of germanium, and this indicates a smaller component of the covalent bond in indium antimonide (apparently both because of the appearance of the ionic component of the bond and because of the enhancement of the metallic component).

The microstructure of polished sections is usually etched with dilute CP-4 or a 5% solution of ferric chloride in dilute (1 : 2) hydrochloric acid[200].

Crystals of the hexagonal modification of the wurtzite type are formed in thin films of indium antimonide[229–232]. For technical applications (e.g., for use as Hall effect detectors) thin films can be prepared by vaporizing bulk antimonide or vaporizing the components separately at different temperatures. The temperature of the substrate should not be lower than 400°C [211, 233]. As the closest electronic and crystallochemical analogue of grey tin, indium antimonide can be used as the seed to convert white tin into grey tin[108].

The technology of preparation is notable for its great simplicity as compared to its analogues, and is the most advanced at the present time. Indium antimonide is obtained by melting stoichiometric amounts of the elements in evacuated and sealed quartz ampoules at a temperature which does not exceed 900°C. The method of recrystallization is very effective for purification.

Listed below are the distribution coefficients of some impurities in indium antimonide according to the data of references[234, 235].

Cu	..	6.6×10^{-4}	Se	..	0·35
Ag	..	4.9×10^{-5}	Te	..	0·8 – 1
Au	..	1.9×10^{-6}	P	..	0·16
Zn	..	2·3 – 10	As	..	5·4
Cd	..	0·86	Fe	..	0·004
Ga	..	2·4	Ni	..	6×10^{-5}
Sn	..	0·057	S	..	0·1
Tl	..	5.2×10^{-4}			

It is apparent that most of the impurities of the acceptor type have distribution coefficients of less than unity, and that impurities of the donor type have coefficients greater than unity. For this reason, in zone recrystallization the first part of the ingot usually possesses hole-type conduction, and the rest of the ingot has electronic conduction.

Indium antimonide can be obtained in the monocrystalline state by using the methods of Bridgman, Czochrapski, and others, but it is more desirable to combine this process with zone recrystallization. The combination of the method of zone refining with the preparation of monocrystalline rods is also significant because, as was explained in reference 236, the distribution coefficient of certain impurities in indium antimonide depends on the crystallographic direction. In ingots with a polycrystalline structure, the individual crystals have different orientations, so that as the zone moves, the impurities are displaced to one side in some parts of the ingot and to the other side in others. Thus, zone recrystallization may lose its effectiveness.

As already mentioned, indium antimonide expands upon crystallizing. As was shown in reference 237, this fact should cause the substance to be carried along in the direction of motion of the molten zone. To prevent this transport, the ingot should be tilted at an angle, as shown in Fig. 20.

Taking all the above into consideration, a monocrystalline substance with the highest carrier mobility ever was prepared by zone melting (in an atmosphere of argon at a pressure of about 1 atm, at a rate of displacement of the zone of 7 to 12 mm/hr, and a number of passes of the zone of 80 to 90)[238].

Indium antimonide is one of the best-studied semiconductors. The number of studies devoted to the investigation of the electrical properties is very large, and increases steadily. This is due to the fact that they are very interesting both from the scientific standpoint and from that of practical application.

Forbidden gap width 0·25 eV
Electron mobility at room temperature[172] 65,000 cm^2/V sec
Hole mobility at room temperature[172] 1,000 cm^2/V sec

Fig. 20 *Apparatus for the zone recrystallization of InSb*

As was stated above, the authors of reference 238 obtained a still higher carrier mobility: at 77°K, it reached values of the order of 1,000,000 cm^2/V sec, i.e., it exceeded the values for any substance known thus far. These properties of indium antimonide (small forbidden gap width combined with an unusually high carrier mobility and certain other parameters) enable one to observe a number of new phenomena in this compound which had not been known earlier and which provide a deeper insight into the nature of the interaction of the atoms in a crystal. Thus, for example, a degenerate state of the "electron gas" can be obtained in the substance at almost any temperature. The electrical conductivity and mobility in such degenerate crystals are strictly constant[239].

In a study of the optical absorption, an anomalous shift of the absorption edge toward short wavelengths was observed as the impurity content increased[240]. There are many other peculiarities in the behaviour of indium antimonide, which are caused by various factors.

In addition to its scientific value, the material has already found broad applications in the manufacture of the following devices: Hall generators, infrared radiation detectors, light filters, etc.[45, 114].

2. Type A^2B^6 Compounds

As was stated above, many compounds of the type A^2B^6 have long been known to chemists; their luminophor properties made them the object of numerous studies. Recently, the interest in these compounds has increased still more because of their use in semiconductor technology. Whereas the luminophors were once considered to be dielectrics, i.e., ionic crystals in which electronic processes can play an important part only under special conditions, today the luminiphors are classed among semiconductors.

Some compounds of type A^2B^6 (e.g., beryllium chalcogenides and magnesium telluride) have been studied very superficially from the standpoint of physicochemical properties, and their electrical properties are not known. On the other hand, cadmium sulphide and other chalcogenides have been the object of a very large number of studies, chiefly in the field of optics and photoelectric properties. Therefore, we are giving below the fullest possible account of the properties of beryllium and magnesium compounds as they are known up to the present time, and also those of zinc oxide, and a brief account of the properties of zinc, cadmium, and mercury chalcogenides.

BERYLLIUM AND MAGNESIUM COMPOUNDS

Beryllium oxide, BeO, is a white powder. It crystallizes in the structure of wurtzite.

Lattice spacings[38] $a = 2.695$ Å, $c = 4.39$ Å
Specific gravity (in the molten state) ~ 3.025 g/cm^3
Melting point $2520 \pm 30°C$
Heat of formation 135 kcal/mole
Moh's hardness[3] 9.0
Index of refraction n_D^{20} 1.73
Solubility in water (30°C) 0.00002 g/100 ml/water

BeO may be obtained by heating beryllium in air or oxygen at temperatures above 800°C. It is a very stable compound which is not reduced by hydrogen, magnesium, sodium, aluminium, etc., and does not react with sulphur. Beryllium carbide, silicide or boride are formed only when beryllium is heated with carbon, silicon or boron. Oxides having other formulae are not known. Beginning at 2300°, it sublimes appreciably. In contrast to most other oxides, it is a good conductor of heat[135].

Beryllium oxide dissolves in concentrated sulphuric acid and in molten potassium hydroxide[69], and is known as a luminophor and also as a catalyst. Descriptions of methods of synthesis and purification specific for luminophors are available; it has scarcely been studied as a semiconductor (dielectric). The forbidden gap width should be very large, of the same order as that of diamond.

Beryllium sulphide is a grey powder. It crystallizes in the structure of sphalerite.

Lattice spacing[241] 4.86 Å
Moh's hardness[3] 7.5
Specific gravity 2.36 g/cm^3
Heat of formation 56.1 kcal/mole

BeS is obtained by the dry method only because water decomposes it to evolve hydrogen sulphide. It is formed as a powder when beryllium is burned in sulphur vapours. It dissolves in acids, evolving hydrogen sulphide. It has been studied as a luminophor[135] but not as a semiconductor.

Beryllium selenide is a dark-brown substance. It crystallizes in the sphalerite lattice with a lattice spacing $a = 5.13$ Å[241]. It has a very unpleasant odour, since it decomposes in moist air with the evolution of hydrogen selenide. The properties of beryllium selenide have hardly been studied at all.

Beryllium telluride is a dark-grey, almost black powder. It crystallizes in the structure of sphalerite with a lattice spacing $a = 5.61$ Å[241].

It decomposes in humid air, which probably accounts for its unpleasant smell. The properties of this compound have been scarcely studied.

Magnesium telluride is a white powder. It crystallizes in the structure of wurtzite with the following lattice spacings:
$$a = 4.52 \text{ Å}; \quad c = 7.33 \text{ Å}[242]$$
It can be obtained by heating magnesium in a stream of hydrogen saturated with tellurium vapours. The white magnesium telluride formed in this reaction turns brown rapidly when in contact with air. The compound is obtained as a dense brown mass when tellurium vapours are passed in vacuum over finely divided magnesium[243]. The heat of formation of magnesium telluride is 50 kcal/mole[51]. It has not been studied as a semiconductor.

ZINC, CADMIUM AND MERCURY COMPOUNDS

Zinc oxide exists as a white powder or transparent crystal, and exists in nature in the form of the mineral zincite. It crystallizes in the structure of wurtzite.

Lattice spacing[48] $a = 3.25 \text{Å}, c = 5.20 \text{Å}$
Specific gravity (exp.) 5.606 g/cm^3
Moh's hardness[3]   5
Index of refraction n_D^{20} 2.01
Melting point $> 1800\,°\text{C}$
Heat of formation[51]   83.17 kcal/mole
Solubility in water at $29\,°\text{C}$[69]   0.00016 g/100 m/water
Forbidden gap width[42] ~ 3.2 eV

The microhardness of zinc oxide has been studied in reference 50.

There exists one more compound of zinc with oxygen, the peroxide, which is used as a medical preparation.

Zinc oxide is obtained in the form of a powder by heating zinc at a temperature close to its boiling point. This very stable compound is reduced by hydrogen with great difficulty. The specific heat, the dissociation pressure as a function of the temperature and the conditions of reduction have been studied in detail in references 152, 244.

Zinc oxide is soluble in acids, alkalis, and ammonium chloride. It is insoluble in ammonia and alcohol.

In order to study its semiconducting properties, pure zinc oxide was prepared both in the form of sintered samples and in the form of single crystals[245]. In the latter, the mobility of the carriers (electrons) varied between 100 and 1000 cm^2/V sec.

This electronic conduction is attributed to the stoichiometric excess of zinc in the interstices [41].

Zinc, cadmium and mercury chalcogenide. The chalcogenides of zinc and cadmium exist in at least two modifications. The character of the polymorphous structures depends on the conditions of the synthesis. Reference 247 gives a brief survey of the work in this field, and a description of the metastable modification of zinc selenide with the alternation of two- and three-layer phases corresponding to hexagonal and cubic packing. A description of the modifications of zinc and cadmium telluride is also given. Polymorphous modifications of zinc chalcogenides have been studied in reference 248.

The structure of thin films of cadmium telluride is described in references 229, 249. Among mercury compounds, only the sulphide has two modifications: the α-modification, which has a special structure, and the β-modification (metacinnabar), with the structure of zinc blende. The selenide and telluride are known only in the structure of zinc blende. The microcleavage of zinc, cadmium and mercury chalcogenides was studied in reference 50, where the results of the investigation formed the basis of a conclusion concerning the magnitude of the component of the ionic bond in these compounds, which is usually greater than the same component in compounds of type A^3B^5.

Certain mechanical properties of zinc and cadmium selenide and cadmium telluride are given in reference 56. The thermal expansion of zinc selenide and cadmium telluride was studied in references 184, 250.

The phase diagrams of most of the cadmium, zinc and mercury chalcogenides indicate the existence of only one compound in each case. The behaviour of these compounds in the molten state has not been sufficiently studied. Only the data of reference 63 are available, where it was found that cadmium telluride in the molten state retains the properties of a semiconductor, and that the fusion of mercury telluride and selenide is associated with complex processes.

The physicochemical properties of zinc and cadmium chalcogenides have been thoroughly studied (the mercury chalcogenides have been studied much less). In reference 251, a comparison was made of the various methods of synthesis available in the literature and the mechanisms of interactions involved in these compounds were investigated, as was the temperature dependence of the saturated vapour pressure of zinc and cadmium tellurides and selenides. Some data obtained from investigations of the microstructure of these compounds and data on the composition of the etchants are given in reference 200.

The basic feature which differentiates the group of zinc and cadmium chalcogenides from compounds of the group A^3B^5, and one which is important in characterizing their semiconducting properties is the following: whereas in compounds of type A^3B^5 the conduction type is determined by foreign impurities, in many compounds of type A^2B^6 a different type of conduction arises from deviations in stoichiometry. This fact is related to the interpretation of compounds of type A^2B^6 as comprising of phases of different compositions for which the thermodynamic conditions of the equilibrium in the crystal-gas system are of paramount importance [203,252].

The effect of foreign impurities on the electrical properties of compounds of type A^2B^6, discussed for example in reference 253, does not follow the simple scheme which was cited for compounds of type A^3B^5, and has not been studied for many of these compounds.

Because of the high melting points, the methods of zone recrystallization and withdrawal of single crystals from the melt are not used in the preparation of zinc and cadmium chalcogenides. An exception is cadmium telluride, for which a method of recrystallization involving the variation of the vapour pressure of the volatile component has been developed, which makes it possible to obtain both electronic and hole-type samples of this substance [252,254]. A more common method involves the preparation of single crystals from the gas phase [255]. Particularly interesting modifications of this method have been developed for zinc sulphide [256] and cadmium sulphide [257,258].

The hydrothermal method of preparing luminophors, known for half a century[259], has been applied in an improved version to semiconductors of type A^2B^6 [260]. The method consists of the treatment of solutions (most often aqueous) at a temperature below their boiling point under a pressure of several hundred atmospheres for long periods.

Mercury chalcogenides are usually obtained by fusion in evacuated and sealed quartz ampoules placed in protective steel vessels. Vibrational stirring is recommended. However, it is difficult to obtain uniform samples, and the excess mercury can be removed only through prolonged annealing [246,261].

As is evident from Table 20, all the chalcogenides of zinc, cadmium and mercury are semiconductors. Zinc and cadmium chalcogenides have a high sensitivity to electromagnetic radiation, visible and ultraviolet light, X-rays and gamma rays, and also corpuscular radiation (α and β). This has made them useful in the manufacture of photoresistances and photocells, photosensitive layers in television camera tubes, in dosimeters, in counters, etc.[42,262]. Mercury selenide and telluride are used for making instruments measuring magnetic field intensities [215].

Table 20 Physicochemical and electric properties of zinc, cadmium and mercury chalcogenides[51, 135, 246–248, 251, 261]

Substance and structure (s — sphalerite, w — wurtzite)	External appearance	Lattice spacings, Å	Specific gravity (exp.) at $20°$, g/cm^3	Microhardness, kg/mm^2	Refractive index n_D^{20}	M.P., °C	Heat of formation, kcal/mole	ΔE (optical), eV	μ_n at $300°C$, $cm^2 V^{-1} sec^{-1}$
ZnS_w	White powder or colourless crystals	$a = 3.811$ $c = 6.234$	4.087	$H_{25} = 178 \pm 27$ in pl. (011) (after Knoop)	2.356	1800–1900 press 100–150 atm	45.3	3.55–3.70	—
ZnS_s	Same	$a = 5.43$	4.102	—	2.368	1020–1024 press 100–200 atm	48.5	3.60–3.64	—
$ZnSe_s$	Same	$a = 5.653$	5.42	$H_{50} = 135 \pm 5$	2.89	>1600	34	2.58–2.66	—
$ZnTe$	Same	$a = 6.087$	6.34	$H_{20} = 90 \pm 5$	3.56	1239	30	2.15	~100
CdS_w	Yellow transparent crystals	$a = 4.14$ $c = 6.72$	4.82	3.2 (after Mohs)	2.506	1750 (press 100 atm)	34.5	2.38–2.48	1460
$CdSe_w$	Dark-grey crystals with reddish tinge	$a = 4.30$ $c = 7.02$	5.81	$H_{20} = 90–130$	—	>1350	25	1.74	—
$CdTe_s$	Black, opaque crystals	$a = 6.46$	6.20 at 15°C	$H_{20} = 60 \pm 5$	2.5	1045	24.3	1.41–1.47	600
HgS_s	Black, opaque crystals	$a = 5.84$	7.73	3 after Mohs	2.85	1750 (press 120 atm) 446 (subl.) 690 ± 10	13.9		>250
$HgSe_s$	Same	$a = 6.074 \pm 0.006$	7.1–8.9	—	—	790	5.1	0.12 (from data electr. measurements)	20000
$HgTe_s$	Same	$a = 6.429 \pm 0.006$	8.17	—	—	670	2–3	0.08 (from data of electr. measurements)	22900

Type A^1B^7 Compounds

COPPER COMPOUNDS

The methods of preparing copper compounds are adequately treated in courses of preparative chemistry; their physicochemical properties are shown in Table 21.

The electrical properties of copper chalcogenides have not been adequately studied. No data are available on copper chloride. Reference 153 gives a forbidden gap width $\Delta E = 2\cdot 94$ eV for copper bromide with a reference to a personal communication.

Table 21 Physicochemical properties of copper chalcogenides[3, 48, 51, 69, 204]

Substance	External appearance	Lattice spacing, Å	Specific gravity g/cm³	Hardness or micro-hardness	M.P., °C	Refractive index	Heat of formation, kcal/mole	Solubility in water, mole/l
CuCl	White powder	$a = 5\cdot 416$	3·53	2·5 after Mohs	422	1·93	32·2	4×10^{-4}
CuBr	ditto	$a = 5\cdot 821$	4·72	$H_{10} = 21\cdot 2$	504	2·116	25·1	7×10^{-5}
CuI	ditto	$a = 6\cdot 053$	5·63	$H_{10} = 19\cdot 2$	605	2·35	16·2	1×10^{-6}

Copper iodide has been studied in the form of thin films[263]. At a film thickness of 180μ, the resistivity was 0·01 ohm and decreased upon illumination. Hevesy[264] holds that copper iodide possesses a mixed electronic-ionic conduction, and that the role of electronic conduction increases as the temperature is lowered. The conductivity of copper iodide depends on the partial pressure of iodine. This problem has been elucidated in reference 265.

SILVER COMPOUNDS

Amongst silver compounds, only silver iodide crystallizes in structures with a tetrahedral arrangement of the atoms. The low-temperature modification of the iodide (light-yellow powder) has the structure of zinc blende (γ). As the temperature rises, silver iodide changes into a hexagonal modification of the wurtzite type (β), and then, at 145·6°C, into a defect structure melting at 552°C (α)[48 p. 192]. The preparation of silver iodide is described in manuals on preparative chemistry. The physicochemical properties of the low-temperature modification as given by references 7, 48, 51, 69, 204 are listed below:

Lattice spacing   6·48 Å
Specific gravity (exp.)   5·67 g/cm^3
Microhardness H_{10} 6·5 kg/mm^2
Index of refraction   2·22
Heat of formation   14·91 kcal/mole
Solubility in water 3×10^{-7} g/100 ml water

The electrical conductivity of silver iodide increases several thousand times when the β modification changes into the α modification. According to the data of Tuband, the electrical conductivity of silver iodide is electrolytic[266]. The conduction appears to be mixed in character, since in AgI the photoelectric effect is observed, whereas Welker reports the presence of a forbidden gap width $\Delta E = 2·8$ eV and an electron mobility $\mu_n = 30$ cm^2/V sec [268].

Part C

Isovalent Solid Solutions

1. SOLID SOLUTIONS BETWEEN COMPOUNDS OF TYPE A^3B^5

STUDIES of solid solutions between the most covalent binary compounds of the crystallochemical group of diamond-like semiconductors have begun quite recently. The first work, carried out in 1953, showed the absence of solid solutions in the system InAs – InSb[269]. However, practical applications required the development of materials with intermediate properties compared to the properties of compounds of type A^3B^5, which had already been studied. The investigations therefore continued. The basis for the work conducted in the Soviet Union involved some theoretical considerations as well.

We believe that the analogy between the compounds of type A^3B^5 and the elements of group IV should extend to the nature of their chemical interaction. By that time, the classical work of Stöhr and Klemm[86], forgotten for over ten years, was already the centre of attention for specialists in semiconducting materials. Studies dealing with the electrical properties of solid solutions of silicon and germanium appeared, which dealt with the entire range of the concentrations obtained by the above-mentioned authors. Thus, investigations of quasi-binary sections of ternary systems based on the compounds A^3B^5 were undertaken.

The studies [270,271] were followed by many others which established the formation of continuous substitutional solid solutions in systems of this type. The technology of preparation of solid solutions proved to be complex. It was found necessary to use prolonged annealing and other methods of homogenization, including the very effective method of zone levelling, which consists of zone melting with the zone travelling alternately in opposite directions[122].

At the time of writing, of all the possible systems involving the replacement of one of the atoms by an atom of the same group of the periodic table, no study has as yet been made of systems involving the participation of nitrides and compounds of boron because of the serious technological difficulties involved in their preparation, and also certain other obstacles. The interactions of the binary components of 18 systems are sketched in Table 22.

Table 22   Ternary systems based on A^3B^5 compounds with isovalent substitution

System	Nature of interaction	Literature sources
AlP—GaP	Not studied	—
AlP—InP	ditto	—
GaP—InP	Nonequilibrium solid solutions over a wide concentration range	272
AlAs—GaAs	Solid solutions	590
AlAs—InAs	Solid solutions over the entire concentration range	158, 273
GaAs—InAs	ditto	270, 274
AlSb—GaSb	ditto	275
AlSb—InSb	ditto	276
GaSb—InSb	ditto	270, 277
AlP—AlAs	Not studied	—
AlP—AlSb	ditto	—
AlAs—AlSb	ditto	—
GaP—GaAs	Solid solutions over the entire concentration range	271
GaP—GaSb	Not studied	—
GaAs—GaSb	Solid solutions over the entire concentration range	278
InP—InAs	ditto	271
InP—InSb	No dissolution was observed	279
InAs—InSb	Solid solutions over the entire concentration range	278

The system GaP – InP. Until recently, no data were available in the literature on the nature of the interaction between the binary components of this system, apparently because of the difficulties involved in its synthesis and homogenization. In reference 280, the authors hold that no solutions are present in this system. However, a synthesis of alloys of this system was recently carried out which made use of vibrational stirring and of excess phosphorus pressure[272]. A study of three alloys revealed all the phenomena which are characteristic of the formation of non-equilibrium solid solutions. The Debye powder patterns showed one system of very broad lines (corresponding to the structure of zinc blende). A study of these patterns showed that the dissolution is more vigorous in the lattice of gallium phosphide than in the lattice of indium phosphide. The microstructure was polyhedral, and no other phases were present. It may be assumed that this system will be obtained in the equilibrium form as the homogenization methods are improved. This is also supported by experience gained in preparing other similar systems, which will be discussed below. The system is interesting from a practical standpoint, since the initial binary compounds possess a number of properties which are of practical value and are peculiar to these substances.

The system AlAs – InAs. The alloys of this system could not be obtained by the usual fusion in evacuated and sealed ampoules because the reaction is incomplete. The aluminium surface becomes covered with a stable film of aluminium arsenide which prevents the diffusion of arsenic into the molten aluminium. The use of vibrational stirring has made it possible to synthesize a number of alloys of this system[273]. A homogeneous state of the alloys was reached only after annealing for 4400 hr at 850°C.

The change of lattice constant with the composition was linear; thermal analysis also confirmed a type of interaction between the binary components which corresponded to the formation of solid solutions. The author of reference 273 holds that this system contains a continuous series of substitutional solid solutions. In the course of a study of the alloys it was noted that in the region of composition located close to indium arsenide, the decomposition in air, usual for aluminium arsenide, does not occur. It was also noted that higher-melting alloys were synthesized and homogenized more easily (rapidly) than low-melting ones.

Practical interest in this system may be elicited in connection with the widespread use of indium arsenide as a material for Hall devices, and with the possibility of varying various associated properties in the alloys of this system (for instance, increasing the forbidden gap width); this may be useful because of the variety of the instrumental applications of the Hall effect.

The system GaAs – InAs was first studied by X-ray diffraction[270]. The X-ray patterns of this system showed broad lines in the form of bands. The positions of the left-hand and right-hand edges of these bands at a composition corresponding to 1 : 1 coincided approximately with the positions of the corresponding bands of the binary components of this system. On the X-ray patterns of alloys of the compositions 1 : 3 and 3 : 1, the broad bands had one more distinct boundary, corresponding to the binary component present in excess.

Before annealing, the alloys were ground into a powder. As the time of the annealing increased, the bands of the Debye powder patterns gradually narrowed, indicating the homogenization of the non-equilibrium solid solutions. This enabled the authors of reference 270 to conclude that solid solutions were formed in this system with a wide range of concentrations, and focused attention on the specific character of their formation, related to the covalent type of attraction between the atoms. This system was later studied in more detail with a larger number of alloys[274]. The change in lattice spacings with composition showed a very close agreement with Vegard's law.

In the next work of these authors[278], the method of X-ray diffraction analysis was used to obtain a solidus line which also corresponded to the formation of a solid solution in the quasi-binary section being studied. A state of equilibrium was reached by annealing the pressed powders at 900°C for two weeks (Fig. 21).

Later, the electrical properties of the alloys of this system and the thermal conductivity were studied in polycrystalline samples obtained by zone levelling[281]. The study also gives a fusibility diagram of the system.

According to the data of reference 281, the forbidden gap width increases smoothly with the gallium arsenide content. The electron mobility decreases in approximately linear fashion up to 70% gallium arsenide. At the same time, the thermal conductivity drops very sharply to a minimum at the composition 1 : 1 as the content of the second component increases. This is explained by the fact that the free path of electrons exceeds the free path of phonons considerably, and the latter are therefore scattered to a much larger extent at various lattice sites (including sites which are statistically substituted by indium and gallium); this corresponds to the observations of A. V. Ioffe and A. F. Ioffe for other solid solutions[227].

In a study published very recently[282], the authors obtained the samples by directed cooling and investigated their optical properties. They showed that the forbidden gap width changed linearly with the composition over an appreciable concentration range.

The system AlSb – GaSb. The existence of solubility was predicted in this system[283] despite the data of reference 284 where no solid solutions were observed.

In a number of subsequent works [275, 285, 286] the study of the interaction between the substances involved the use of thermal analysis, the investigation of microstructure and microhardness, and investigation of the distribution of the electrical conductivity over the ingot. Because of the closely similar lattice spacings of the compounds AlSb and GaSb, the X-ray method was not used in the first work, and the formation of solid solutions was established later through the use of a precision method. For the homogenization of the alloys of this system (the first to be reported for solid solutions of binary compounds), the authors of these studies used zone levelling. The heater was moved alternately in opposite directions at the rate of 9 to 10 mm/hr. After the 20th pass of the zone, the middle part of the ingot was subjected to analysis.

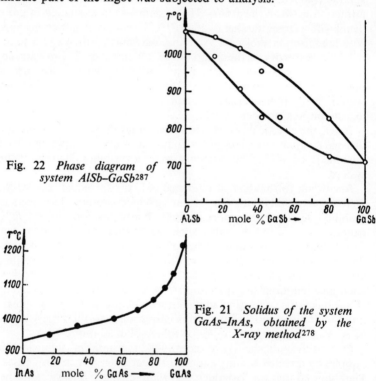

Fig. 22 Phase diagram of system AlSb–GaSb[287]

Fig. 21 Solidus of the system GaAs–InAs, obtained by the X-ray method[278]

A chemical analysis was performed in order to follow the chemical composition of the alloys after the zone levelling[287]. A study established the formation of substitutional solid solutions along the quasi-binary section of AlSb – GaSb. The fusibility diagram (Fig. 22) has a shape characteristic of all substitutional solid solutions of type I according to Roozeboom.

The corrosion resistance of the alloys of the system is much greater than that of aluminium antimonide. Investigations were carried out at 21 °C in an atmosphere of saturated water vapour and in liquid media (water, 25% sodium hydroxide). At a concentration of gallium antimonide in the alloys in excess of 50%, all the media used had no appreciable effect on these alloys even when the experiment lasted a long time. This curious fact, first observed with alloys of this system, was later observed with alloys of another system based on indium antimonide (see below), in which an increase in the corrosion resistance was also obtained.

Homogeneous, coarsely crystalline ingots were used to measure the temperature dependence of the electrical conductivity and of the Hall effect, and also the optical absorption. It was found that the temperature dependence of the electrical properties of the solid solutions under study was similar to the dependences in the initial binary compounds. This is of great interest from the standpoint of the possibility of a practical application and classification of solid solutions of A^3B^5 in the same crystallochemical group of semiconductors; this is discussed in more detail below.

As was shown in reference 286, when the composition of the solid solution changes, the forbidden gap width changes smoothly between 0·7 and 1·6 eV, and the mobility of the carriers (holes) changes from 150 to 525 cm^2/V sec.

Reference 288 confirmed the existence of solid solutions in this system. Monocrystalline samples obtained by directed cooling and zone recrystallization showed an approximately linear variation of the lattice spacing with the composition and also an approximately linear variation of the forbidden gap width with a change in the lattice spacing.

These studies made it possible to obtain equilibrium alloys of this system in a completely polycrystalline and monocrystalline state, with given properties within the limits of the properties of the initial compounds. This is undoubtedly of great interest in technology. The most appropriate area of application of semiconducting materials based on alloys of the system is obviously the production of rectifiers and of photocells[285].

The system AlSb – InSb. Preliminary investigations enabled the authors to state that this system contains a continuous series of substitutional solid solutions[283]. Later[276, 289, 290] homogeneous samples of the alloys were obtained.

It was found that the homogenization of the alloys involves serious difficulties. Zone levelling, prolonged annealing, and annealing under pressure did not produce any positive results. This may be explained by the fact that the difference in the melting points of the constituent compounds is almost twice as great as the differences in the cases of alloys of other systems. The only method which permitted a relatively fast homogenization of the samples of alloys was the one in which they were annealed in the form of a powder. However, the impossibility of accurately determining the electrical properties of matter in the powdered state made it necessary to improve on this method.

An effective method was developed which involved tempering and subsequent annealing at temperatures above those corresponding to the non-equilibrium solidus curves[291]. The first step of the method consisted of the use of high rates of cooling for the given system (quenching of the liquid alloy in a solution of table salt, which promoted the removal of the vapour jacket from the object undergoing the quenching). This process led to an extensive heterogenization of the alloys, as determined from the data of a microstructural investigation and X-ray analysis. The area of contact of the interacting phases having various concentrations of the initial binary components was increased considerably, thus promoting the fusion during the second step, which consisted of a prolonged annealing.

It should be noted that in the case of other substances, the first step (quenching) can itself give rise to the formation of equilibrium or quasi-equilibrium structures[292]. This means that the temperature gradient during the quenching of these substances is sufficient to create conditions of diffusionless crystallization[293], something which was not observed in the present case.

The alloys of the system AlSb – InSb, following synthesis (corundum crucibles are recommended) with vigorous vibrational stirring (at 1200°C for one hour) were tempered in a solution of table salt. The dense, finely crystalline ingots obtained were annealed at temperatures established by means of a thermal analysis of the non-equilibrium alloys. These temperatures corresponded approximately to the middle of the temperature range in which the first thermal effect was observed during the differential thermal analysis, and were always above the melting point of the low-melting component of the system. Under these conditions of annealing, no liquefaction of the alloys was observed, and the samples were homogenized for 120 to 500 hr. The microhardness of the alloys following the homogenization changed along a curve with a maximum at about 90 mole % aluminium antimonide. The lattice spacings changed linearly with the composition. A study of the Debye powder patterns showed that the dissolution of indium in the lattice of aluminium antimonide proceeds faster than does the dissolution of aluminium in the lattice of indium antimonide.

A phase diagram of the system, based on the data of reference 290, is shown in Fig. 23. A calculation of the heat of mixing made by Kamenetskaya[294] showed that in solid solutions of this system there is a tendency toward the combination of like components of the alloy[295]. The decomposition should take place at temperatures below 450°C. However, the small diffusion rates in the solid phase insured the stability of the solid solutions, as was observed in the experiments. The homogeneous solid solutions obtained in the system under study showed a considerable increase in the corrosion resistance as compared with aluminium antimonide; this was also observed in the case of alloys of the system AlSb – GaSb.

It should be noted that the authors of reference 278 were not able to obtain homogeneous alloys in the central region of compositions (approximately from 40 to 80 mole % aluminium antimonide) apparently because of a bad choice of homogenization conditions (annealing for three months at 525°C). Nevertheless, the authors also believe that continuous solubility is present in this system. They give an equilibrium phase diagram (solidus) based on an X-ray investigation of the alloys as powders.

The electrical properties were successfully investigated in samples obtained by a specially developed homogenization method[291], with a charge carrier concentration $n \approx 3 \times 10^{17}$ cm^{-3} [296,297]. It was found that the forbidden gap width and the carrier mobility (all the samples had hole conduction) changed smoothly with the composition, giving values intermediate between the values of these parameters in the initial binary components. As reported by the authors [297], certain characteristics of the change in the electrical parameters with the temperature make it possible to assume that the samples were solid solutions and not mixtures (even very finely dispersed ones) of binary components.

A study of the mechanism of carrier scattering involving an investigation of the Ettingshauser-Nernst thermomagnetic effect led to the important deduction that the disordered structure of a solid solution, which results from a statistical substitution of aluminium by indium, gives rise to a very minor amount of scattering in this system. The electronic processes in a solid solution have the same general character as those in the binary components of the system.

In addition to the theoretical significance of this conclusion, which will be discussed below, it follows from it that the carrier mobility in the alloys of this system is limited by foreign impurities and may be increased by purifying the material. Thus, one may hope that materials based on these solid solutions will find practical applications in the same areas where aluminium antimonide is now used, and that they will possibly be even more promising.

The system GaSb – InSb. The formation of solid solutions in this system was first established by X-ray diffraction[270]. The shape of the Debye powder patterns following synthesis, and their variation with the time of the annealing of the alloys, was the same as in the GaAs – InAs system described above. However, in contrast to the latter, the homogenization of the antimonide proceeded much more rapidly. The results of the investigation enabled the authors to conclude that continuous substitutional solid solutions were formed. Simultaneously with this work the article[284] was published, in which the results of a thermal analysis led to the conclusion that a eutectic interaction took place in this and also in other similar systems. Later[298], after repeating the work, the authors concurred with the conclusion of reference 270.

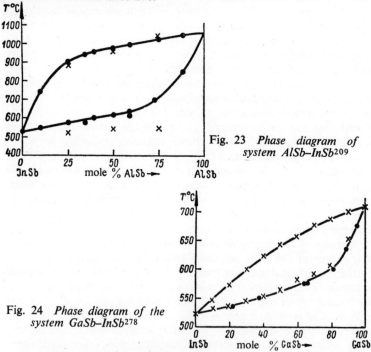

Fig. 23 *Phase diagram of system AlSb–InSb*[209]

Fig. 24 *Phase diagram of the system GaSb–InSb*[278]

For this reason, the next step taken in the study of this system was the thermal analysis of alloys[283]. The phase diagram corresponding to type I solid solutions, according to Roozeboom, was elucidated. The same type of phase diagram was obtained by using the X-ray method to establish the solidus line in reference 278. Fig. 24 shows the latter phase diagram.

It was shown that a phase diagram of eutectic shape was obtained because of the non-equilibrium state of the alloys of this system. The authors of reference 284 did not take into account the characteristics of the crystallization and diffusion of the substitutional solid solutions of compounds with covalent bonds. Earlier, crystallization under conditions of slow diffusion in the solid phase had been discussed, and it was shown that a non-equilibrium mixture of solid solutions of various compositions is obtained in this case, and that the phase diagram is similar to that of a degenerate eutectic [90].

An equilibrium state of the system is attained only when the diffusion in the solid phase is able to equalize the concentration in the portions of the alloy which separate out. This condition was fulfilled in reference 270, where the substances were annealed in the form of powders for long periods of time. In reference 277, the X-ray method, applied to powders, established that the diffusion of indium into the lattice of gallium antimonide proceeds faster than the diffusion of gallium into the lattice of indium antimonide. These conclusions were supported in reference 298.

Annealing in the powdered state, a method which is very convenient for the solution of fundamental problems concerning the nature of the interaction of the components, which was also used in later work[278], ceased to be satisfactory as soon as it became necessary to study the electrical and optical properties of alloys.

Attempts to study these properties in poorly homogenized samples have led to a repetition of the erroneous assertion that solid solutions were absent from this system[299,300]. Therefore, further efforts were directed towards obtaining alloys of this system not only in the homogeneous state but also in the coarsely crystalline and monocrystalline states. Two methods were employed for this purpose: zone levelling and directed cooling. Homogeneous alloys with a coarsely crystalline structure and a practically constant composition along the length of the ingot were obtained by the first method, which was suggested in reference 301 and developed for germanium-silicon alloys[302], by decreasing the rate of travel of the zone and thus approaching the equilibrium condition of crystallization. The carrier concentration reached a record value for solid solutions of $n = 2 \times 10^{15}$ cm^{-3}.

Studies of the electrical and optical properties of these samples gave the following results[303]. The forbidden gap width increases monotonically with the composition as the content of gallium antimonide increases. The electron mobility decreases smoothly with this change in composition, and this dependence is almost linear. At the middle point of the compositions, the value of the mobility is of the order of 30,000 cm^2/V sec. All this indicates that, in the general mechanism of electron scattering, the disordered structure must play a minor role. It was also established that the substitution of gallium for indium in this system leaves the hole mobility practically unchanged and is chiefly manifested in a change of the electron mobility. The alloys of this system retain the characteristic properties of compounds of type A^3B^5 and behave in a manner analogous to these compounds.

By studying the character of the interaction of the components in the solid solutions of this system by means of a thermodynamic analysis of the phase diagram[295], one can conclude that the small value of the energy of mixing and the small diffusion rate permit one to expect the solid solutions in the system GaSb – InSb to be stable below the solidus temperature[303]. In reference 295, the heat of mixing was determined from the best agreement between the experimental curves and curves calculated by assuming regular solutions[294]. The authors developed a method for the numerical calculation of the liquidus and solidus of binary systems.

Using directed cooling, another method for the homogenization of solid solutions, the authors of reference 304, having studied the electrical properties of such samples of alloys, came to conclusions which were somewhat different from the preceding ones. The forbidden gap width determined by studying the optical absorption of the alloys, also changes smoothly with the composition (giving a curve with a slight depression). However, the electron mobility drops sharply as the content of gallium antimonide increases to 30 mole%, and then remains almost unchanged up to pure gallium antimonide.

It is possible that the discrepancy of the results is due to a somewhat inferior homogenization of the alloys in the method of directed cooling. There are several aspects to the possibility of practical application of materials based on this system. First, it is possible that some alloys will be able to replace indium arsenide, since they possess approximately the same properties but do not contain arsenic, which is not without certain dangers in manufacture. Secondly, the alloys of this system may be used to develop semiconductor devices with a variable forbidden gap width[305].

The system GaP – GaAs was first studied in reference 271. The author concludes that continuous solid solutions are formed in this system. Apparently, the synthesis proceeds somewhat more easily than in other systems of this type, but is also accompanied by the formation of non-equilibrium solid solutions.

Crystals of the solid solutions gradually change their external appearance when the concentration of gallium arsenide changes. For the compositions Ga($As_y P_{1-y}$), or $0 < y < 0.1$, the crystals are transparent and orange in colour. At $0.1 < y < 0.3$, the colour becomes reddish, at $y \approx 0.4$, red, at $y \approx 0.5$ dark red, and at $y > 0.6$, the crystals become opaque. The variation of the forbidden gap width with the composition is not linear for the system, and is reminiscent of the change of this parameter in the silicon-germanium system. In the latter, as was mentioned above, the relation is approximated by a broken line with two almost linear segments. When a small amount of silicon is added, the forbidden gap width increases sharply, and undergoes little change when the composition changes from 20 mole% to 100 mole% silicon.

This type of dependence is explained in reference 94 by the difference in the structure of the energy bands in silicon and germanium. The authors of reference 172 come to the same conclusion by considering the dependence of the forbidden gap on the composition in the system GaP – GaAs.

The system GaAs – GaSb has not been sufficiently studied. On the one hand, it is believed[280] that solid solutions are absent from this system. On the other hand[278], on the basis of preliminary experiments it is reported that solid solutions may exist over the entire concentration range; the same view is held in reference 306.

The system InP – InAs was first studied in reference 271. Later, these results were confirmed in reference 307 and the same system was studied in reference 298. The authors of the latter found, as did the author of reference 271, that the lattice spacings change linearly with the composition. According to their observations, ordinary synthesis without additional annealing gave equilibrium alloys. Fig. 25 shows the phase diagram of the system according to the data of reference 298.

Reference 271 reported a linear change of the lattice spacing in the system as a function of the composition (in conformity with Vegard's law) and concluded that substitutional solid solutions were formed over the entire concentration range. Also reported were data on the electrical properties, which were later studied in more detail[308]. The author found that the forbidden gap width of the system varies linearly with the composition, indicating a similarity in the structures of the energy spectrum of the initial dynamic compounds. The carrier mobility increases monotonically, and at indium arsenide concentrations in excess of 60 mole% reaches a value of the order of 10,000 cm^2/V sec.

The temperature dependence of the electrical parameters of the alloys of this system is similar to this dependence in indium arsenide. With regard to the introduction of impurities and the formation of hole or electronic conduction by these impurities, the solid solutions of this system behave just as do compounds of type A^3B^5.

The optical[309] and thermoelectric[116] properties of the alloys were also studied. The authors of the latter work came to the conclusion that the solid solutions of this system are promising material for use in devices generating thermoelectromotive force.

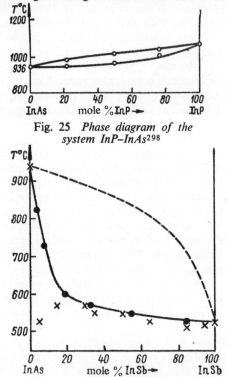

Fig. 25 *Phase diagram of the system InP–InAs*[298]

— — — — Shih and Peretti data; X — one-phase region; ● — X-ray analysis data

Fig. 26 *Phase diagram of the system InAs–InSb*[278]

A study has also been published[310] in which the X-ray diffraction method was used to evaluate the intracrystalline liquefaction, following zone levelling, from the broadening of the diffraction lines. The maximum liquefaction was observed in the range of 10 to 50 mole % indium phosphide. Alloys of this solid solution are conveniently prepared by a method developed for two or more volatile components[311].

These alloys found practical applications in the manufacture of devices for measuring magnetic field intensities[312].

The system InAs – InSb was first studied in reference 269 and then in reference 270; in both studies, no solid solutions were found in the systems (at least in amounts exceeding 3 mole %).

In 1958 the formation of substitutional solid solutions over the entire concentration range of this system was reported in reference 278. Serious difficulties were noted in the homogenization of the alloys. After 3 months of annealing at 525°C, the alloys still had two phases. Only when careful annealing was carried out for several weeks at temperatures from 525 to 600°C, depending upon composition, were solid solutions obtained over the entire concentration range. The authors obtained the solidus curve by the X-ray method.

The phase diagram as given by reference 278 is shown in Fig. 26.

Reference 313 also confirms the formation of continuous substitutional solid solutions in this system. In addition, it concludes that indium antimonide dissolves more easily in indium arsenide. The authors recommend the method of zone levelling for the homogenization.

The electrical properties of this system have not yet been studied, despite the fact that it combines the properties of the most interesting substances of type A^3B^5 (from the standpoint of high carrier mobility and photosensitivity in the infrared region).

Isovalent solid solutions A^3B^5 with the substitution of one component by two components have not been studied. Only in reference 298 was a hypothesis advanced concerning the existence of wide solubility regions in the system AlSb – GaSb – InSb.

Solid solutions with the simultaneous replacement of both components of A^3B^5 have not been studied; only in reference 279 is a partial solubility noted in the system InP – GaSb.

2. Solid Solutions Between Compounds of Type A^2B^6

Solid solutions between certain compounds of type A^2B^6 have long been known as natural minerals[314] and luminophors[135]. Lately, interest has been expressed in them because of their semiconducting properties and the possibility of obtaining intermediate combinations of properties as compared to the properties of the initial compounds.

Table 23 gives a brief description of the interaction of the components in solid solutions based on the compounds A^2B^6. No data are available in the literature concerning the formation of solid solutions by compounds of beryllium and magnesium.

In those investigated, structural transformations are observed in certain cases (for example, from the structure of zinc blende into that of wurtzite in the systems ZnSe – CdSe and HgS – CdS); this is of considerable scientific interest.

Table 23 Ternary systems based on type A^2B^6 compounds with isovalent substitution (w *wurtzite structure*, s *sphalerite structure*)

Systems	Nature of interaction	Literature sources
ZnS—CdS	Continuous solid solutions	135, 315
ZnS—HgS	Continuous solid solutions (s)	316
CdS—HgS	Solid solutions with a transfer $w \to s$ in the region of 43 to 60 mole % HgS	316
ZnSe—CdSe	Solid solutions with a transfer $s \to w$ with ~50 mole % CdSe	317, 318
ZnSe—HgSe	Continuous solid solutions	314
CdSe—HgSe	Not studied	—
ZnTe—CdTe	Continuous solid solutions	319 *et al*
ZnTe—HgTe	ditto	320
CdTe—HgTe	ditto	320, 321, 322
ZnS—ZnSe	ditto	323
ZnS—ZnTe	Solid solutions from the binary components aspect (~5–10 mole %)	324
ZnSe—ZnTe	Continuous solid solutions	323
CdS—CdSe	Continuous solid solutions (*w*)	257, 325, 326
CdS—CdTe	Not studied	—
CdSe—CdTe	ditto	—
HgS—HgSe	Continuous solid solutions (*s*)	314, 327
HgS—HgTe	ditto	327
HgSe—HgTe	ditto	246, 327

The methods used for preparation are the same as those used for the preparation of binary compounds of type A^2B^6. For solid solutions based on sulphides, the hydrothermal method has been used in addition to preparation from the gas phase. Compositions which are difficult to prepare, such as solutions of zinc and mercury sulphide, have been synthesized by this method[316]. The method found most suitable for tellurides is that of fusion of the components in sealed quartz ampoules[320-322].

Insofar as can be judged from data available in the literature, the problems of homogenization of alloys are of paramount importance in the case of tellurides, and can be solved by means of prolonged annealing[320]. For the lowest-melting compositions, the method of zone levelling will apparently be as effective as for solid solutions based on compounds of type A^3B^5.

The electrical properties of the system investigated show them to be semiconducting materials with a high sensitivity to light and other radiations. In solid solutions involving mercury chalcogenides, very high carrier mobilities are observed[246]. The variation of the forbidden gap width may be both linear or close to linear (the systems ZnS – ZnSe, CdS – CdSe), and non-linear with a gently sloping minimum (the system ZnSe – ZnTe).

Solid solutions A^2B^6 will apparently find practical applications in the same areas as binary compounds. Of particular interest may be the combination of the high carrier mobility with a high sensitivity to radiation.

Isovalent solid solutions with the substitution of one component by two have not been studied. Such compositions exist among minerals, as do products of the simultaneous substitution of both components, for example, Hg, Zn (S, Se) – guadalcazarite[314], etc.

Quaternary alloys of ZnSe – CdTe have been synthesized, but a study of their electrical properties showed an incomplete homogenization of the alloys[117].

The interaction of zinc chalcogenides with manganese chalcogenides was also studied[202]. Solid solutions with transformations of the structures from tetrahedral (s, w) to octahedral (NaCl) are formed in these systems. Certain alloys have a conductivity with an order of magnitude typical of semiconductors.

3. Solid Solutions Between Compounds of Type A^1B^7

Compounds of type A^1B^7 with a sphalerite structure form continuous substitutional solid solutions. According to the old work[328], the systems CuCl – CuBr and CuBr – CuI are characterized by phase diagrams of type I in the Roozeboom classification.

According to more recent data[329], systems of this type give fusibility diagrams with a minimum. In the systems CuCl – CuBr, CuBr – CuI, and AgI – CuI, the minimum is located at temperatures of 408, 443 and 500°C, respectively.

Part D
Ternary and More Complete Heterovalent Phases
1. Ternary Phases

As was already stated in the second section of the first chapter, of the ten possible types of nondefect ternary tetrahedral phases, representatives of five "dicationic types" have been identified. Let us consider these types in more detail.

PHASES OF TYPE $A^1B^3C_2^6$

Chemical compounds of this type were synthesized and investigated by X-ray diffraction in reference 330. The authors synthesized alloys involving copper and silver (A), aluminium, gallium, indium, thallium (B), and sulphur, selenium, and tellurium (C).

The alloys were synthesized and sealed in evacuated quartz ampoules which were held at temperatures from 600 to 1000°C; they were kept above the melting point for 12 hours and below the melting point for 8 days.

All the synthesized alloys had a structure similar to that of chalcopyrite. The authors regard the latter as a cubic close-packed lattice of chalcogenide atoms in the tetrahedral vacancies of which the metal atoms are ordered. Thus, the chalcopyrite structure is very close to that of zinc blende and differs from the latter by a certain tetragonal character, i.e. by a compression of the lattice in the direction 'c'. This compression arises from the ordered arrangement of the atoms A^1 and B^3 (Fig. 27).

The process of formation of ternary compounds of this type is considered in reference 330 as a dissolution of compounds of the univalent metal in the compound of the trivalent metal and as the formation of a phase with superlattice structure in the region of solid solutions.

A significant region of homogeneity was indeed observed in these compounds – those in which the ionic radii of the two metals were similar. In addition, these regions of homogeneity were always observed on the side of the trivalent component and never exceeded the stoichimetric composition when the univalent component (i.e. the chalcogenide of the univalent metal) was in excess. In compounds of metals with very different radii, as in the case of silver – gallium compounds, for example, the region of homogeneity is limited only by the stoichiometric composition. These phenomena can be adequately interpreted from the standpoint of formation of ternary tetrahedral phases, which was discussed in Chapter 1.

Later, some properties of compounds of this type were studied: the melting point, the microhardness, the forbidden gap width and, in a few cases, the thermo-emf, the carrier mobilities, etc. [16, 331-333, 591]. All the measurements were carried out on polycrystalline and sometimes coarsely crystalline samples (these substances have not yet been obtained in the monocrystalline form).

Table 24 lists the basic properties of investigated compounds of this type. In it the compounds are arranged in order of increasing atomic weight, which leads to an increase of the lattice constant. The coefficients of linear expansion measured for some of the substances in this group were found to be of the same order as those of compounds of type A^3B^5 [331]. The electron mobility in one of them

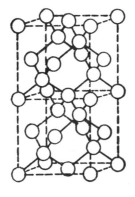

Fig. 27 *Structure of chalcopyrite*

($CuInSe_2$) was found to be 1150 cm²/V sec, which does not confirm the hypothesis put forward in reference 332 that high mobilities are scarcely possible in compounds with a chalcopyrite structure. Compounds of this type possess photoconductivity and exhibit rectification on point contacts. The area of their practical application has not yet been defined. Compounds of type $A^1B^3C_2^6$ are close analogues of compounds of type A^2B^6 so it is possible that their practical application will be in the same area.

It was found that chalcopyrite, $CuFeS_2$, whose name was applied to a whole structural type, is itself a semiconductor. This fact suggested the synthesis of its analogues, where sulphur would be replaced by other chalcogenides and silver would be substituted for copper[331]. However, the structure of these compounds was found to be more complex, although the semiconducting properties were

Table 24 Some properties of compounds of type $A^1B^3C_2^6$

Compound	Lattice spacing a, Å	Lattice spacing ratio, c/a	Microhardness, kg/mm^2	M.P., °C	ΔE, eV
$CuAlS_2$	5·31	1·96			
$CuGaS_2$	5·34	1·96			
$CuInS_2$	5·51	2·00			1·2
$CuTlS_2$	5·58	2·00			
$CuAlSe_2$	5·60	1·94			
$CuGaSe_2$	5·60	1·96	430	1040	1·63
$CuInSe_2$	5·77	2·00	260	990	0·92
$CuTlSe_2$	5·83	1·99	90	405	1·07
$CuAlTe_2$	5·96	1·97			
$CuGaTe_2$	5·99	1·99	360	870	1·0
$CuInTe_2$	6·16	2·00	210	700	0·95
$AgAlS_2$	5·69	1·80			
$AgGaS_2$	5·74	1·79			
$AgInS_2$	5·81	1·92			1·9
$AgAlSe_2$	5·95	1·80			
$AgGaSe_2$	5·97	1·82	450	850	1·66
$AgInSe_2$	6·09	1·92	230	773	1·18
$AgAlTe_2$	6·29	1·88			
$AgGaTe_2$	6·29	1·90	180	720	1·1
$AgInTe_2$	6·40	1·96	190	675	0·96

retained. (The electron mobility in the compound $AgFeTe_2$ reached high values. At carrier concentrations of the order of 10^{18} cm^{-3}, it exceeded 2000 cm^2/V sec, which makes this substance interesting from the practical standpoint.) Thus, whereas manganese sulphide and selenide were exceptions among binary tetrahedral phases, the chalcopyrite, containing a different element, iron, was an exception among the ternary phases. The replacement of element B in this type by elements of group V (antimony or bismuth) leads to the formation of substances with the structure of rock salt (their description – they are also semiconductors – exceeds the scope of this discussion). However, as was established in reference 334, when arsenic participates in such compositions as $CuAsS_2$ and $CuAsSe_2$, a structure arises which is very similar to that of sphalerite. These compounds have not been studied in more detail.

PHASES OF TYPE $A^2B^4C_2^5$

Chemical compounds of this type were studied in references 335–338. They have the structure of chalcopyrite as do compounds of type $A^1B^3C_2^6$. Only two of them, obtained to date, $ZnSnAs_2$ and $MgGeP_2$, have the structure of zinc blende. The same group of substances includes $BeSiN_2$ with the wurtzite structure[339].

Table 25 Some properties of compounds of type $A^2B^4C_2^5$

Compound	Structure	Lattice spacing, Å	Lattice spacing ratio, c/a	ΔE, eV
$BeSiN_2$	Wurtzite	$2 \cdot 87_2$	$1 \cdot 62_7$	
$MgGeP_2$	Sphalerite	$5 \cdot 65_2$	—	
$ZnSiP_2$	Chalcopyrite	$5 \cdot 39_8$	$1 \cdot 93_4$	
$ZnSiAs_2$	Chalcopyrite	$5 \cdot 60_8$	$1 \cdot 94_2$	2·1
$ZnGeP_2$	Chalcopyrite	5·46	1·97	2·2
$ZnGeAs_2$	Chalcopyrite	5·67	$1 \cdot 96_7$	>0·6
$ZnSnP_2$*	Chalcopyrite			2.1
$ZnSnAs_2$	Sphalerite	$5 \cdot 85_1$		
$CdSiP_2$	Chalcopyrite			
$CdSiAs_2$	Chalcopyrite			
$CdGeP_2$	Chalcopyrite	$5 \cdot 73_8$	$1 \cdot 87_8$	1·8
$CdGeAs_2$	Chalcopyrite	$5 \cdot 94_2$	$1 \cdot 88_9$	
$CdSnP_2$*	Chalcopyrite			1·5
$CdSnAs_2$	Chalcopyrite	$6 \cdot 09_2$	$1 \cdot 95_7$	0·26

*Obtained as two-phase substances

Table 25 lists some properties of these substances.

Their synthesis, except that of the nitride, is carried out in the usual manner in evacuated and sealed quartz ampoules. Their physicochemical properties are largely unknown. It was found relatively recently that one of these substances, $ZnGeAs_2$, forms substitutional solid solutions with germanium, and in the process assumes the structure of zinc blende[340]. It may be safely assumed that this property of giving regions of homogeneity with an element of group IV is typical of many of the compounds of type $A^1B^4C_2^5$. From the general standpoint of formation of tetrahedral phases, the problem of the width of the homogeneity region of these compounds along sections connecting defect and excess binary phases is of particular interest. However, the study of this problem is still in its initial stages.

It has been established that the compound $CdSnSb_2$ does not exist in the pure state [20, 21]. Attempts to synthesize it lead to the formation of a mixture of substances, CdSb and SnSb; this has been shown very definitely by X-ray diffraction and microstructural investigations[23].

Thus, the most difficult substance in this group is the compound $CdSnAs_2$. In connection with the search for semiconductors with high carrier mobilities and also because of the fact that this compound should have a lower melting point than the rest, it has been the one most closely studied.

Whilst the other substances were obtained only in the form of polycrystals, and information about them is limited to the data of Table 25, $CdSnAs_2$ has been obtained in a coarsely crystalline[341] and also in a monocrystalline form[342].

Measurements made on single crystals have shown that this compound has a smaller forbidden gap width than indium arsenide (its electronic analogue), and the electron mobility in $CdSnAs_2$, despite the participation of three components in the formation of the tetrahedral phase, lies within the same range as that of indium arsenide. The authors of reference 342 believe that the mobility in the ternary compound may exceed that in the binary one. Given below are the data on the electrical properties of this compound according to reference 342;

Forbidden gap width   0·26eV

Electron mobility at a carrier concentration of 1×10^{17} cm^{-3}, 22,000 cm^2/V sec.

Investigations have shown that the carrier mobility of $CdSnAs_2$ is limited by the presence of foreign impurities and may be increased by purifying the material. Studies of the properties of $CdSnAs_2$ showed, for the first time, that a carrier mobility may exist in ternary compounds which is of the same order as in binary compounds. Moreover, such a high mobility was first observed in a substance with a chalcopyrite structure, and this is also very significant.

All of the above suggests that the compound $CdSnAs_2$ and probaby many other compounds of this type will be promising for practical applications in electronic devices.

PHASES OF TYPE $A^1B_2^4C_3^5$

This type of tetrahedral phase has many representatives. The first such compound with a zinc blende structure, i.e. with a statistical distribution of the copper and germanium atoms in the lattice points of zinc blende, $CuGe_2P_3$, was found by the authors of references 279, 343. In subsequent work it was observed that most of the analogues of this type do not exist in the pure state [22].

The substance $AgGe_2P_3$ was found to be a chemical compound with a structure that has not yet been definitely identified. Later a report appeared which confirmed the existence of the chemical compound $CuGe_2P_3$ having the structure of sphalerite. It also gave data on one more substance of this type, $CuSi_2P_3$, which has the same structure[338].

Table 26 lists only a few data on the properties of this type of compound[279, 338, 343].

Table 26 Properties of compounds of type $A^1B_2^4C_3^5$

Compound	Structure	Lattice spacing, Å	Melting point, °C	Microhardness H_{50}, kg/mm²
$CuSi_2P_3$	Sphalerite	5·25	?	—
$CuGe_2P_3$	Sphalerite	5·37$_5$	840	850 ± 20
$AgGe_2P_3$	?	—	742	615 ± 30

No complete physicochemical study of the system copper – germanium – phosphorus has been made. However, the predicted compound $CuGe_2P_3$ was identified. Similarly, on the basis of the general principles of formation of tetrahedral phases (see Chapter I) the formation of solid solutions of this compound with germanium was predicted and identified[340] over a wide concentration range (~ 30 mole %). The nature of the replacement of the atoms in $CuGe_2P_3$ by germanium was investigated in reference 21. Reference 343 established the solubility of this compound in hypothetical Ge_3P_4, which is of interest from the standpoint of formation of defect tetrahedral phases. The electrical properties of these compounds have not yet been studied.

PHASES OF TYPE $A_2^1B^4C_3^6$

All the known compounds of this type crystallize in the structure of zinc blende or one similar to it.

Table 27 lists the available data on their structure as given by references 16, 35, 344, 389.

Table 27 Structure of compounds of type $A^1B^4C_3^6$

Compound	Structure	Lattice spacings, Å	
		a	c
Cu_2SiTe_3	Sphalerite	—	—
Cu_2GeS_3	Diamond-like	5·32	10·41
Cu_2GeSe_3	Diamond-like	5·58	10·96
Cu_2GeTe_3	Diamond-like	5·92	11·85
Cu_2SnSe_3	Sphalerite	5·70	—
Cu_2SnTe_3	Sphalerite	6·04	—

The physicochemical and electrical properties of these compounds have not been studied at all. The very latest work[592], however, gives some electrical properties.

PHASES OF TYPE $A_3^1B^5C_4^6$

Compounds of this type contain an element of group V, which acts as a "cation". Minerals of type $A_3^1B^5C_4^6$ are found in nature, such as enargite Cu_3AsS_4, famatinite Cu_3SbS_4, and the mineral Cu_3PS_4, which have orthorhombic, cubic, and orthorhombic lattices respectively. Substances synthesized by analogy with these minerals have been studied very little. In patent 345 it is pointed out that the forbidden gap width of Cu_3AsS_4 and Cu_3SbS_4 is of the order of 0·8 eV. Table 28 shows a few known data for these compounds.

Table 28 Properties of compounds of type $A_3^1B^5C_4^6$

Compound	Crystal lattice	Melting point, °C
Cu_3AsS_4	Orthorhombic	655
Cu_3AsSe_4	Orthorhombic	—
Cu_3SbS_4	Cubic	555
Cu_3SbSe_4	Cubic	425

In a recent study, electron diffraction was used to investigate the structure and phase composition of Cu_3SbS_4 in thin films[346].

The remaining five types of ternary phases belong to "monocationic" compounds, for which only fragmentary data exist. The author of reference 16 reports that Al_2GeSe, Al_2GeTe and Al_2SnTe, which he synthesized, were completely unstable, and the more stable In_2GeSe and In_2GeTe crystallize in very complex structures. In reference 22 it was noted that when compounds of type $A_2^3B^4C^6$ are synthesized by the fusion of the initial components in evacuated and sealed ampoules, the reaction usually proceeds in the direction of the formation of the defect phase $A_2^3C_3^6$.

In reference 22, an attempt was made to obtain solid solutions of compounds of type A^3B^5 and $A_2^3B^4C^6$; the attempt was not successful.

Quaternary and More Complex Phases

As was explained in Section 2, Chapter 1, all four-component phases should have a variable composition, since they represent a combination of simpler binary and ternary compounds.

Compared to that of ternary compounds, the probability of formation of broad homogeneity regions in quaternary phases increases considerably. As was evident from the discussion of ternary phases, the latter are combinations of defect and excess compounds with respect to sphalerite. The middle point of the compositions is the one where the most favourable combination arises for the formation of a chemical compound with an electron concentration different from that of the components. In quaternary compositions, however, the phases constituting them are equivalent in characterisitics, so that the possibility of thier dissolution in one another increases. Nevertheless, under special conditions, the formation of compounds with a narrow homogeneity aegion is also possible in quaternary alloys. This may be the case, for example, when, owing to a large difference in the parameters of the constituting atoms, conditions are created for ordering, i.e. for the formation of a super-lattice.

Thus, the mineral stannite, Cu_2FeSnS, exists in nature, and there also exist artificially prepared analogues, $Cu_2FeSnSe_4$, Cu_2FeGeS, $Cu_2FeGeSe_4$, and $Cu_2NiGeSe_4$, all of diamond-like structure[16], which can be regarded as super-lattice formations in solid solutions of the ternary compound $A_2^1B^4C_3^6$ plus A^2B^6 (A^2 is in this case divalent iron and nickel). However, most of the quaternary heterovalent compositions studied to date have been phases of variable composition. (It may be noted that according to the data of reference 347, a mixture of tellurides is formed instead of the compounds $Cu_2CdSnTe_4$, etc.)

On the basis of the classification given in Section 2, Chapter 1, certain quaternary systems may, for convenience, be treated as derivatives of compounds of type A^3B^5, and certain others as derivatives of compounds of type A^2B^6, and a part of the quaternary systems (the most complex ones) as derivatives of various ternary compounds. These types include quaternary systems which are combinations of compounds A^2B^6 and A^3B^5. Being the simplest they are of considerable interest; they also make it possible to obtain intermediate properties between compounds which have already been thoroughly investigated (A^2B^6) and those which have found practical application as semiconductors (A^3B^5).

SYSTEMS BASED ON COMPOUNDS OF TYPE A^3B^5

As early as 1955, on the basis of the fact that the binary compounds ZnSe and GaAs (individually) gave continuous solid solutions with the same compound Ga_2Se_3, it was suggested that heterovalent solid solutions could form in this, the system GaAs – ZnSe [4], and then it was shown experimentally that this system contains continuous substitutional solid solutions[348]. Subsequently, research along these lines developed further.

Table 29 gives a brief description of the interaction of the components in quaternary systems of this type.

Table 29 Quaternary heterovalent systems based on compounds A^3B^5 and A^2B^6

System	Nature of interaction of binary components	Reference
AlP — ZnS	Insignificant solubility of AlP in ZnS, less than 1 mole %	349
AlSb — CdTe	Continuous solid solutions (s)	—
GaAs — ZnSe	ditto	348
InP — CdS	Solid solutions over a wide concentration range (w)	279
InP — CdSe	Solid solutions over a wide concentration range (s)	279
InP — CdTe	Solubility not recorded	279
InAs — CdTe	Solubility 30 mole % (from the indium arsenide end) (s)	350
InAs — HgTe	Continuous solid solutions (s)	351
InSb — CdTe	Solid solutions in the range of 95 to 100 mole % In Sb (s)	352

It is necessary to add that in the majority of the systems, or at any rate in all the systems where this was investigated, the region characterized by the absence of solid solutions is usually a eutectic, a mechanical mixture of the initial components with a solid solution, or a mixture of two solid solutions. As should have been expected, the formation of a chemical compound was not observed in any of these cases. If this formation is possible, it is apparently so only in systems including elements of the very first periods (for more detail see Chapter 3).

Table 30 Quaternary heterovalent systems based on compounds of type A^3B^5 and on ternary compounds

System	Nature of interaction of binary compounds	Reference
GaAs — ZnGeAs$_2$	Nonequilibrium solid solutions over a wide concentration range (?)	353
InAs — ZnGeAs$_2$	Continuous solid solutions (s)	353, 354
InAs — ZnSnAs$_2$	ditto	353, 350
InAs — CdSnAs$_2$	Continuous solid solutions (s – chalcopyrite)	20, 353
InSb — ZnSnSb$_2$*	Solid solutions from 100 to 20 mole % InSb (s)	354
InSb — CdSnSb$_2$*	Solid solutions from 100 to 50 mole % InSb (s)	20, 23

*Hypothetical compound

The physicochemical and electrical properties of these systems have thus far been scarcely studied at all. The data shown in Table 29 were obtained by the X-ray method.

Systems based on compounds A^3B^5 and ternary compounds are also beginning to be studied. Table 30 lists data on the interaction of the components in these systems.

In reference 353, while developing the ideas of the synthesis of many-component systems, the author proposes eight-component systems of the general formula

where

$$[(A^2_{xy/2} B^2_{x/2\,(1-y)} C^3_{(1-x)z} D^3_{(1-x)\,(1-z)}) (E^4_{xt/2} F^4_{x/2\,(1-t)})] [G^5_u H^5_{1-u}]$$

$$\text{at } 0 \leqslant \begin{Bmatrix} x \\ y \\ z \\ t \\ u \end{Bmatrix} \leqslant 1.$$

It might be observed in this connection that the possibility of the formation of many complex systems based on an isovalent substitution is indisputable; it is certainly broader than the possibility of a heterovalent substitution of atoms which is limited by seven combinations (from the number of valence groups of the periodic system). The proposed type of system comprises only four kinds of atoms of different valences.

The electrical properties of the systems enumerated in Table 30 have been scarcely studied at all. The only system for which this study has been made is $InAs - CdSnAs_2$ [355-357]. However, very pure samples could not be obtained with alloys of this system and therefore the interpretation of the results of electrical measurements was not definite.

In reference 354 a study was also made of a system based on indium arsenide and a compound with a chalcopyrite structure of the group $A^1B^3C^6_2$, $InAs \cdot CuInTe_2$. Substitutional solid solutions were found which had the structure of sphalerite over the entire concentration range. The variation of the lattice spacing with the composition obeyed Vegard's law.

In reference 23, a study was undertaken to determine the possibility of the formation of solid solutions based in InSb and of all existing and hypothetical "dicationic" ternary tetrahedral phases of this isoelectronic series. The fusion method was used for the synthesis.

The results were interesting, for it was possible for the first time to obtain series of homogeneous phases, which were solid solutions of types of compounds which have not, so far, been studied.

The formation of homogeneous phases with the structure of sphalerite based in InSb, of the hypothetical compounds $AgSn_2Sb_3$, Ag_2SnTe_3 and of the compound $AgInTe_2$ was proved possible (the system Ag_2SnTe_3 – InSb is quinary).

The interaction in the system $AgInTe_2$ – InSb was studied more thoroughly. It was established that the structure of chalcopyrite, in which the compound $AgInTe_2$ crystallizes, becomes disordered upon dissolution in indium antimonide, that in alloys a statistical distribution of two kinds of atoms (Ag and In) is achieved at the sites corresponding to the positions of the "cations," and that two kinds of atoms (Sb and Te) are statistically distributed at the sites of the "anions" in the sphalerite lattice.

The lattice spacing of sphalerite in the alloys of this system changes as a function of the alloy composition along a straight line connecting the values of a_{InSb} and

The phase diagram of the alloys of this system belongs to Roozeboom's third type. The existence of a region of decomposition of solid solutions has been established in the range of concentrations from 80 to 50 mole % $AgInTe_2$. Measurements of the electrical parameters of the alloys of the system showed that they are all typical semiconductors.

Of particular interest is the system $InSb \cdot Ag_3SbTe_4$, where the formation of solid solutions has been established in a small concentration range (from 100 to 75 mole % InSb). They differ from the preceding ones in the fact that indium and antimony are apparently located in the crystallographically equivalent positions of the "cations", something which is not observed in binary compounds. This fact once again confirms the idea expressed in reference 19 that instances are possible in which the elements of group IV or V of the periodic system act as "cations" and "anions" simultaneously.

Most recently, solid solutions were discovered in the region of concentrations corresponding to a low content of $ZnSiP_2$ [24] in the pseudoternary system GaP – Si – $ZnSiP_2$ [24]. The method of synthesis used by the author involved crystallization from excess gallium. The ternary compound could not be obtained separately. The solubility of gallium phosphide in silicon was observed only in the presence of zinc; this caused the pseudoternary system mentioned in reference 24 to be investigated. The authors explain the formation of phases in this system from the standpoint of the balancing of charges.

SYSTEMS BASED ON COMPOUNDS OF TYPE A^2B^6

Two similar systems, $ZnS - CuGaS_2$ and $ZnS - AgGaS_2$ were studied in reference 358. The authors studied zinc sulphide as a luminophor and discussed the data for the system from this point of view. The synthesis was carried out in evacuated and sealed quartz ampoules held for 12 hr at 900°C. Under these conditions, the author observed the solubility of $CuGaS_2$ in ZnS over the entire concentration range, and the solubility of $AgGaS_2$ in ZnS up to 5 – 10 mole %. In the system $ZnS - CuGaS_2$, the ZnS crystals had the sphalerite structure; from 100 to 60 mole %, in the system $ZnS - AgGaS_2$, this structure was preserved only up to 95 mole % ZnS.

In reference 354, a study was made of the system $CdTe - CuInTe_2$, in which substitutional solid solutions with a zinc blende structure were found over the entire concentration range.

Finally, solid solutions were observed in reference 359 over the entire concentration range (but only after heat treatment) in the systems $CdTe - AgInTe_2$ and $HgTe - AgInTe_2$, and also in the system $CdTe - CuInTe_2$ [360].

The authors also report that they were able to achieve solubility in the quaternary system $CdTe \cdot HgTe \cdot AgInTe_2$*. Electrical measurements on alloys of these systems show that low carrier concentrations can be obtained (of the order of 10^{16} cm^{-3}). The carrier mobility varies over a wide range, decreasing as the composition becomes more complex in the pseudoternary system.

SYSTEMS BASED ON TERNARY PHASES

Thus far, only one such quaternary system has been studied; it is located in the isoelectronic series of germanium, $CuGe_2As_3 - Cu_2GeSe_3$ [361]. The first component of this system, $CuGe_2As_3$, is a hypothetical compound. Investigations have shown that there is a one-phase region with the sphalerite structure in the range of compositions from 60 to 80 mole % Cu_2GeSe_3. The microhardness remained almost unchanged in the one-phase region ($H_{50} = 430$ kg/mm²). The lattice spacing changed from 5·560 to 5·557 Å.

The alloy whose composition corresponded to the middle point of the system was subjected to zone recrystallization at a rate of zone travel of 0·5 to 1·5 cm/hr and a number of passes from 7 to 10. The second phase, which crystallized at the end of the ingot, was

---
*The fifth component is isovalent.

thus completely separated. Physicochemical analysis established the homogeneity of the composition, the preservation of the sphalerite structure and other properties; the ingot was dense and coarsely crystalline. Data on the zone recrystallization of the alloy of this system are of interest as the first successful attempt at zone recrystallization of a quaternary system.

QUINARY AND SEXTUPLE SYSTEMS

In addition to studying the above-mentioned quinary systems Ag_2SnTe_3 – InSb, there is evidence for the formation of a homogeneous alloy at the middle point of the system $CuGePSe$ – InP [279]. Most recently, several alloys of sextuple heterovalent systems of the isoelectronic series of germanium have been obtained in the author's laboratory.

The alloys whose composition corresponded to the general formulae:

$$Cu_{16}Zn_{33}Ga_{113}Ge_{17}As_{131}Se_{48},$$
$$Cu_{57}Zn_{33}Ga_{243}Ge_{51}As_{288}Se_{96},$$
$$Cu_{19}Zn_{99}Ga_{225}Ge_{41}As_{288}Se_{96}, \text{ etc.}$$

contained a small amount of a second phase, found by microscopic analysis. X-ray investigations of all three alloys showed only one system of sharp lines corresponding to the structure of sphalerite with a lattice spacing $a = 5.65$ Å. The ingots had a dendritic structure.

Results obtained from the zone recrystallization of alloys of the system $CuGe_2As_3$ – Cu_2GeSe_3 permit one to expect a successful homogenization of these complex compositions as well, for the purpose of a thorough investigation of their physicochemical and electrical properties.

Such complex systems are very interesting because they should have low thermal conductivities and high electrical conductivities (see Chapter 3). Such a combination of properties may make these materials interesting for use in thermoelectric systems.

Part E
Defect Tetrahedral Phases
1. Binary Compounds with a Defect Structure

ACCORDING to what was stated above (see Chapter 1, Section 3), in considering defect tetrahedral phases one can visualize the existence of tetrahedral phases with electron concentrations from 4·57 to

6·0 el/at. However, among binary phases of the defect type which have the sphalerite structure, representatives of only one type, $A_2^3B_3^6$, are known, and will be described below.

After Hahn and Klingler[362] established that the compounds Ga_2S_3, Ga_2Se_3, Ga_2Te_3 and In_2Te_3 crystallize in the zinc blende structure, where the atoms of gallium or indium occupy only 2/3 of the sites available to them, experiments in 1954 elucidated the relation of compounds of this type to nondefect substances of the same structure[4].

The following conclusion was reached on the basis of the formation of solid solutions over a wide concentration range between gallium selenide and gallium arsenide and zinc selenide. Compounds of type $A_2^3B_3^6$ are similar to the compounds A^2B^6 and A^3B^5 not only in structure but also in bond type (in the character of the density distribution), and hence they should be semiconductors.

It was suggested that these compounds be regarded as members of the isoelectronic series of the diamond crystallochemical group which are located between the compounds of type A^2B^6 and A^3B^5. In the author's view[4], compounds of type $A_2^3B_3^6$ should occupy a position intermediate, in forbidden gap width, between the corresponding isoelectronic compounds A^3B^5 and A^2B^6 in such isoelectronic series. The conductivity of substances of type $A_2^3B_3^6$ should be lower than that of the corresponding compounds A^3B^5.

The photoelectric properties of all substances of type $A_2^3B_3^6$ were studied in reference 363. The assumptions made in reference 4 were confirmed both qualitatively (all the compounds with a defect tetrahedral structure were found to be semiconductors) and quantitatively (their forbidden gap width lay between the values of the neighbouring compounds A^2B^6 and A^3B^5 in the corresponding isoelectronic series). At the present time, compounds of type $A_2^3B_3^6$ are being investigated in various countries.

Table 31 lists the structural types in which substances of the general formula $A_2^3B_3^6$ crystallize.

As is evident from Table 31, the chalcogenides of aluminium and gallium crystallize in tetrahedral structures. Of the indium compounds, only the telluride crystallizes in the structure of sphalerite. However, the β-modification of indium selenide is very similar in structure to zinc blende[366]. The cleavage of two compounds of this type was studied in reference 49.

The two compounds which were studied, Ga_2Te_3 and In_2Te_3, showed a very pronounced cleavage along (111). Reference 367 suggests a mechanism for the formation of a chemical bond in In_2Te_3 (and other compounds of type $A_2^3B_3^6$) based on considerations

Table 31 Structural types of compounds $A_2^3 B_3^6$ [48, 64-366]

Elements	O_3	S_3	Se_3	Te_3
B_2	B_2O_3	?	—	—
Al_2	α, Al_2O_3 β, w γ, Fe_2O_3	α, w^{**} β, w γ, Al_2O_3	w^{**} w	?
Ga_2	α, Al_2O_3 β, monoclinic, stable	s^* α, w^{**} β, w^* > 600° C γ, ZnS < 550° C	s^*	s^*
In_2	Tl_2O_3	α, γ, Al_2O_3 β, spinel, stable	α, In_2Se_3 β, In_2Se_3 > 200° C (6-layered, hexagonal) γ, ?	β, s^* α s^{**} (low-temperature)
Tl_2	Mn_2O_3	—	—	—

*Arbitrary designations: w^* – defect wurtzite structure; w^{**} – wurtzite structure with ordered arrangement of vacancies; s^* – defect sphalerite structure; s^{**} – sphalerite structure with ordered arrangement of atoms (and vacancies) in the structure.

of the hybridization of electrons and resonance of the states of tellurium atoms. It is thought that the free electron pairs of tellurium atoms which surround a vacant site form a bond with this site, thus producing a gain in energy.

The authors of reference 367 estimate the ionic character of the bonds in In_2Te_3 in per cent, using electronegativity values. A confirmation of the relatively low value of the ionic character in In_2Te_3 is considered to be that there is only a small difference between the static and the optical permittivities.

The phase diagrams of the systems Al – S, Al – Se, Al – Te, Ga – Te, In – S, In – Te are given in reference 61. The phase diagrams of the systems Ga – S, Ga – Se, In – Se have not been studied; only individual compounds of these systems have been investigated. In systems of this type there are two or more compounds. Structures with a tetrahedral arrangement of the atoms are encountered only in compounds of type $A_2^3 B_3^6$.

Table 32 shows the melting points of gallium and indium chalcogenides (AB and $A_2^3 B_3^6$).

As can be seen from the table, the melting points of tellurides of

Table 32 Melting points of gallium and indium, °C

Elements	O	S	Se	Te	Elements	O_3	S_3	Se_3	Te_3
Al	—	1200	—	—	Al_2	—	1100	~953	~895
Ga	—	965	960	824	Ga_2	~1900	~1250	>1020	790
In	—	692	660	696	In_2	~2000	1050	890	667

type AB exceed the melting points of tellurides of type $A_2^3B_3^6$ (cf., for example, InTe – In_2Te_3, GaTe – Ga_2Te_3). In gallium and indium sulphides and selenides, the situation is reversed. This may be important in determining the character of the interaction in more complex systems.

The methods of synthesis used for preparing these compounds are the usual ones. Aluminium sulphide is obtained by the action of sulphur on aluminium powder at 700 to 1000°C for a period of 6 to 8 hr[368].

Aluminium selenide is obtained by precipitation from a salt solution, but, like aluminium telluride, is obtained more often by the fusion of the components in graphite crucibles placed in quartz ampoules[369]. Gallium and indium chalcogenides are obtained by fusing the components, preferably with the use of vibrational stirring[158]. Good results were obtained with the method of directional freezing in the preparation of Ga_2Te_3 [370].

Table 33 lists certain physicochemical and electrical properties of semiconductors of type $A_2^3B_3^6$ with a defect tetrahedral structure (for the most-studied modifications).

The heat capacity of gallium chalcogenides in the temperature range of 25 to 100°C was studied in reference 372, where the heat of formation of Ga_2Se_3 was also determined and found to be 110 ± 5 kcal/mole.

The electrical properties of compounds of type $A_2^3B_3^6$ [363, 366, 367, 371] are of great interest, since these compounds have a definite and large number of "cationic" structural vacancies (5.5×10^{21} per cm^3), whose role in the scattering of charge carriers should be important. This problem was thoroughly investigated in reference 367, where In_2Te_3 was selected as the substance to be studied.

The investigation showed that the principal mechanism of scattering of electron waves in In_2Te_3 is the scattering by cationic vacancies. When In_2Te_3 becomes ordered, the carrier mobility decreases, and the thermal conductivity of the lattice increases[365] as a result of a decrease in the free-path length of the electrons and an increase in the free-path length of the phonons. The wavelengths of an electron is ten times as large as that of a phonon, so that the increase in the distance between the scattering lattice points makes the latter very effective scatterers of electron waves and decreases the probability of phonon scattering.

Table 33 Physicochemical and electric properties of compounds of type $A_2^3 B_3^6$ [48, 204, 366, 373]

Compound	Colour	Specific gravity (exp.), g/cm³	Microhardness under 20 g load, kg/mm²	Lattice spacing Å	Thermal conductivity at 300°K, $\frac{kcal}{cm\ sec\ deg}$	ΔE at 300°K (opt. meas.) eV	Electron mobility at 300°C, cm²/V sec
α-Al₂S₃	White	2.32		$a = 6.423 \pm 0.003$ $c = 17.83 \pm 0.02$		4.1	
Al₂Se₃	"					3.1	
Al₂Te₃	Deep yellow					2.5	
β-Ga₂S₃	Light yellow	3.63	500±18	$a = 3.678$ $c = 6.016$		2.5	
Ga₂Se₃	Red	4.92	316±13	$a = 5.41$	0.00122	1.9	10
Ga₂Te₃	Black	5.57	237±9	$a = 5.87_4$	0.00112	1.35	50
β-In₂Se₃	"	5.48	30—50	$a = 4.02$ $c = 19.2$		1.25	125
α-In₂Te₃	"	5.79		$a = 18.50$	0.00268	1.12 at 0°K	2—10
β-In₂Te₃	"	5.73	166±7	$a = 6.14$	0.00166	1.026	50

It was found that already at $-100°C$, In₂Te₃ possesses intrinsic conduction. The introduction of impurity atoms to an amount of 1 atomic % did not, in general, lead to the appearance of extrinsic conduction or to a change in the carrier concentration (with the exception of the impurities bismuth and iodine).

The authors explain these facts on the basis of the arguments developed in reference 374, to the effect that the magnitude of the additional electric field arising in the crystal lattice, as a result of departures from periodicity in the arrangement of the atoms, exceeds the activation energy of the impurity centres, and for this reason the formation of donors and acceptors becomes impossible. This similarity in the behaviour of defect and amorphous or vitreous semiconductors enables the authors of reference 367 to regard defect semiconductors as an intermediate link between normal crystals and amorphous substances.

A study of the electrical properties of Ga_2Te_3 containing small amounts of copper gave the following results: it was found that an admixture of copper of 10^{-6} to 10^{-1} atomic % (with respect to the number of gallium atoms) affects the properties of the substance considerably. The lattice spacing increases by 2·7%, the forbidden gap width decreases, and the substance ceases to be a semiconductor. Thus far, no simple explanations have been offered for this phenomenon[375].

In reference 376, while studying the electrical properties of Ga_2Se_3, the authors found that hole conduction predominates in gallium selenide. In order to explain the constancy of the Hall constant and of the differential thermo-emf in the temperature range up to 1000°K, where these values approach zero, the authors suggested that the hole mobility in gallium selenide was greater than the electron mobility. This differentiates the investigated substance sharply from its non-defect analogues of type A^3B^5, where the ratio of the electron mobility to the hole mobility is always the reverse.

The low thermal conductivities obtained for all compounds with defect structures make them, and the more complex phases based on them, interesting subjects to be studied from the standpoint of possible applications in the field of thermoelectricity.

Studies of defect compounds from the scientific point of view are very interesting, since they make it possible to follow the influence of defects and of their ordering on the physicochemical and electrical properties of substances. The group of substances of this type can be extended considerably to include the identified substitutional solid solutions: $Ga_2Se_3 - Ga_2Te_3$, $Ga_2Te_3 - In_2Te_3$, $In_2Se_3 - In_2Te_3$[397, 377, 378]. In the first system, an ordering of the type of chalcopyrite was observed.

Ternary Tetrahedral Heterovalent Defect Phases

As has been stated earlier, binary defect tetrahedral phases which actually exist crystallize only in the type $A_2^3B_3^6$. Let us now consider a group of ternary defect tetrahedral phases which are analogues of binary defect compounds with the lowest electron concentration, 4·57 el/at. Despite the fact that binary defect phases of this type have not been identified individually, there exist representatives of at least four types of ternary analogues of these substances (see Table 13).

PHASES OF TYPE $A_2^1B^2C_4^7$

Representatives of this type of compounds have been studied in references 33, 379. They include the chemical compounds Cu_2HgI_4 and Ag_2HgI_4, the high-temperature modifications of which crystallize in a structure similar to that of sphalerite with one quarter of the statistically distributed valences in the cationic part of the lattice. The low-temperature modifications also crystallize in tetrahedral (tetragonal) structures which differ from one another to a certain extent.

β-Cu_2HgI_4 crystallizes in a lattice with space group D_{2d}^{11} – 142m, and β-Ag_2HgI_4 crystallizes in the space group $S_4^2 - 1\overline{4}$; in the first substance the metal atoms are arranged in layers, and in the second, more uniformly in the tetrahedral vacancies of the cubic close packing of iodine atoms. Similar results were obtained in reference 380 from an X-ray study of these substances. The latter form solid solutions with one another which crystallize in the structure of β-Cu_2HgI_4 with a statistical distribution of the copper and silver atoms over the copper sites[372, 381].

Table 34 lists certain data for these substances.

Table 34 Physicochemical properties of compounds of type $A_2^1B^2C_4^7$

Substance	Colour	Lattice spacings, Å	Specific gravity, g/cm³
α-Cu_2HgI_4	Brown	6.103 ± 0.005	6·009
α-Ag_2HgI_4	Red	6.383 ± 0.005	5·930
β-Cu_2HgI_4	Red	$a = 6.09 \pm 0.005$ $c = 12.22_8 - 0.00_1$	—
β-Ag_2HgI_4	Yellow	$a = 6.31_4 \pm 0.005$ $c = 12.62_8 \pm 0.00_9$	—

Both of these compounds were first obtained chemically by the coprecipitation of AgI and HgI_2. They were later synthesized by fusing the elements in an evacuated quartz ampoule with subsequent annealing lasting several days. The transition temperature of the modifications ($\alpha - \beta$) is 70°C for the copper compounds and 50°C for the silver compounds. The electrical properties of these compounds have not been studied.

PHASES OF TYPE $A^2 B_2^3 C_4^6$

This type of ternary tetrahedral phase was identified and studied by X-ray diffraction in references 6, 382, 383.

Table 35 gives the formulae of the substance studied.

Table 35 Phases of type $A^2 B_2^3 C_4^6$ with a tetrahedral structure and a spinel structure*

Elements	Zn	Cd	Hg	Elements
Al	$ZnAl_2O_4$ * $ZnAl_2S_4$ $ZnAl_2Se_4$ $ZnAl_2Te_4$	$CdAl_2O_4$ * $CdAl_2S_4$ $CdAl_2Se_4$ $CdAl_2Te_4$	— $HgAl_2S_4$ $HgAl_2Se_4$ $HgAl_2Te_4$	O S Se Te
Ga	$ZnGa_2O_4$ * $ZnGa_2S_4$ $ZnGa_2Se_4$ $ZnGa_2Te_4$	$CdGa_2O_4$ * $CdGa_2S_4$ $CdGa_2Se_4$ $CdGa_2Te_4$	— $HgGa_2S_4$ $HgGa_2Se_4$ $HgGa_2Te_4$	O S Se Te
In	 $ZnIn_2Se_4$ $ZnIn_2Te_4$	$CdIn_2O_4$ * $CdIn_2S_4$ * $CdIn_2Se_4$ $CdIn_2Te_4$	— $HgIn_2S_4$ * $HgIn_2Se_4$ $HgIn_2Te_4$	O S Se Te

**Asterisk denotes substances crystallizing in the spinel structure; the remaining substances crystallize in tetrahedral structures. In the compound $ZnAl_2S_4$, the tetrahedral (wurtzite) structure appears only above 1000°C*

The authors of reference 6 hold that the spinel structure of certain compounds of Table 35 is formed because the octahedral co-ordination is favoured in the corresponding binary compounds (oxides or sulphides).

Table 36 gives the structural types of the tetrahedral phases.

The structures of the ternary phases studied are treated in reference 6 as superlattices in the region of solid solutions of the binary compounds A^2B^6 and $A_2^3B_3^6$.

The structures of the phases $CdAl_2S_4$, $HgAl_2S_4$, $CdAl_2Se_4$, Hg_2AlSe_4, $CdGa_2S_4$, $HgGa_2S_4$, $ZnGa_2Se_4$, $CdGa_2Se_4$, $HgGa_2Se_4$, $ZnGa_2Te_4$, $CdGa_2Te_4$, $HgGa_2Te_4$, $ZnIn_2Se_4$, $HgIn_2Se_4$, $ZnIn_2Te_4$, $CdIn_2Te_4$, $HgIn_2Te_4$ correspond to the space group S_4^2–14 (tetragonal unit cell with two-formula units). This derives from the chalcopyrite structure, in which two sites in the cationic lattice are unoccupied. In the substances $ZnGa_2S_4$, $ZnGa_2Se_4$, $ZnGa_2Te_4$, because of the identical scattering power of the zinc and gallium atoms, one should not differentiate between the space group S_4^2 and the more complex group D_{2d}^{11} – $1\bar{4}2m$.

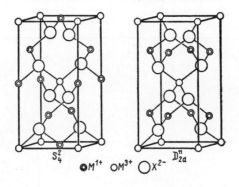

Fig. 28 *Structure of S_4^2 and D_{2d}^{11}*[6]

In addition to the space groups S_4^2 and D_{2d}^{11}, the substances $ZnAl_2Te_4$, $CdAl_2Te_4$, $HgAl_2Te_4$, can also be referred to the space groups D_{2d}^{11} (tetragonal pseudocubic structure with one molecule in the unit cell) and D_{2d}^9 (statistical distribution of the atoms of both metals in a tetragonally distorted lattice of ZnS) (Fig. 28). The compound $CdIn_2Se_4$ belongs to the space group D_{2d}^{11} – $p\bar{4}2m$. The substance $HgGa_2Te_4$ is apparently a solid solution between binary

Table 36 Data of X-ray diffraction study of compounds of type $A^2B_2^3C_4^6$

Substance	Colour	Lattice spacings, Å			Specific gravity, g/cm³		Space groups or structural type
		a	c	c/a	exp	calc.	
$ZnAl_2S_4$	Light grey	$3,76_3$	$6,14_2$	$1,64_2$	$2,6_5$	$2,72_7$	w
$CdAl_2S_4$	Brown	$5,56_4$	$10,3_2$	$1,82_0$	$3,0_4$	$3,06_2$	S_4^2
$HgAl_2S_4$	Black	$5,48_8$	$10,2_6$	$1,87_3$	$4,0_8$	$4,11_2$	S_4^2
$ZnAl_2Se_4$	"	$5,50_3$	$10,9_0$	$1,98_2$	$4,3_7$	$4,37_6$	S_4^2
$CdAl_2Se_4$	"	$5,74_6$	$10,6_8$	$1,85_9$	$4,5_0$	$4,54_2$	S_4^2
$HgAl_2Se_4$	"	$5,70_7$	$10,7_4$	$1,88_2$	$5,0_2$	$5,05_3$	S_4^2
$ZnAl_2Te_4$	"	$5,10_4$	$12,0_5$	$2,03_8$	$4,9_1$	$4,95_5$	$S_4^2 D_{2d}^{11}$
		$5,10_4$	$6,02_7$	$1,01_9$			$D_{2d}^1 D_{2d}^9$
$CdAl_2Te_4$	"	$6,01_1$	$12,2_1$	$2,03_0$	$5,1_0$	$5,09_9$	$S_4^2 D_{2d}^{11}$
		$6,01_1$	$6,10_7$	$1,01_5$			$D_{2d}^1 D_{2d}^9$
$HgAl_2Te_4$	"	$6,00_4$	$12,1_1$	$2,01_7$	$5,7_9$	$5,81_6$	$S_4^2 D_{2d}^{11}$
		$6,00_4$	$6,05_7$	$1,00_9$			$D_d^1 D_{2u}^9$
$ZnGa_2S_4$	White	$5,27_3$	$10,4_4$	$1,97_9$	$3,7_5$	$3,80_8$	$S_4^2 D_{2d}^{11}$
$CdGa_2S_4$	Yellowish brown	$5,57_7$	$10,0_8$	$1,80_8$	$3,9_7$	$4,03_2$	S_4^2
$HgGa_2S_4$	Yellowish green	$5,50_7$	$10,23$	$1,86_0$	$4,9_5$	$5,00_2$	S_4^2
$ZnGa_2Se_4$	Yellowish brown	$5,49_6$	$10,9_9$	$2,00$	$5,1_3$	$5,21_5$	$S_4^2 D_{2d}^{11}$
$CdGa_2Se_4$	Dark green	$5,74_2$	$10,73$	$1,87_0$	$6,2_8$	$5,32_6$	S_4^2
$HgGa_2Se_4$	Black	$5,71_4$	$10,7_8$	$1,88_6$	$6,1_0$	$6,18_5$	S_4^2
$ZnGa_2Te_4$	"	$5,93_7$	$11,8_7$	$2,00$	$5,5_7$	$5,67_4$	$S_4^2 D_{2d}^{11}$
$CdGa_2Te_4$	"	$6,09_3$	$11,8_1$	$1,93_8$	$5,6_3$	$5,77_1$	S_4^2
$HgGa_2Te_4$	"	$6,01_7$	$12,0_3$	$2,00$	$6,4_2$	$6,48_1$	$S_4^2 D_{2d}^{11}$
		$6,01_7$	—	—			ZnS
$ZnIn_2Se_4$	"	$5,71_0$	$11,42$	$2,00$	$5,3_6$	$5,44_3$	S_4^2
$CdIn_2Se_4$	"	$5,81_7$	$5,81_7$	$1,00$	$5,5_4$	$5,54_8$	D_{2d}^{11}
$HgIn_2Se_4$	"	$5,76_3$	$11,8_0$	$2,04_7$	$6,2_6$	$6,33_1$	S_4^2
$ZnIn_2Te_4$	"	$6,12_2$	$12,2_4$	$2,00$	$5,8_2$	$5,82_6$	S_4^2
$CdIn_2Te_4$	"	$6,20_4$	$12,4_0$	$2,00$	$5,8_8$	$5,92_4$	S_4^2
$HgIn_2Te_4$	"	$6,18_6$	$12,3_7$	$2,00$	$6,3_4$	$5,59_5$	S_4^2

components with a ZnS structure.

At the basis of all the structures presented (to describe the phases of type $A^2B_2^3C_4^6$) lies the nearly cubic close packing of chalcogenide atoms, in the tetrahedral vacancies of which metal atoms are distributed in accordance with the space groups assigned to the given phases.

The authors of reference 6 propose to refer to the structure in which most of the investigated substances crystallize (space group S_4^2) as the "thiogallate" structure, since it was first observed in the compound $CdGa_2S_4$.

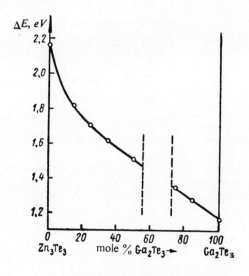

Fig. 29 *Forbidden gap width ΔE versus the composition of the system $ZnTe$–Ga_2Te_3* [370]

The width of the homogeneity region was studied only in the eight substances, shown in Table 36. It may be noted, however, that the homogeneity regions of ternary phases can be large for a minor distortion of the structure ($c/a \simeq 2 \cdot 00$) and are usually small when the distortion is pronounced. This follows from the results of reference 383, where in the two systems Ga_2Se_3 – ZnSe and Ga_2Te_3 – ZnTe only substitutional solid solutions with the sphalerite structure were observed over a wide concentration range. These are the phases whose lattice spacing ratio c/a is equal to 2·00.

It is interesting that certain ternary phases of type $A_2^2B_2^3C_4^6$ have proved more inert to acids and other chemical agents than the corresponding initial binary compounds. According to the observations of the authors of reference 6, the heat stability falls from the sulphides to the tellurides.

When compounds with different metal components were compared, the compounds containing aluminium or mercury were found to be the most difficult to synthesize. As was shown in further investigations, the synthesis of these is conveniently carried out by the usual fusion of the elemental components in sealed and evacuated quartz ampoules. When a mixture of the initial binary components is pressed and heated, the preparations obtained are partly fused and partly sintered[6] and are not suitable for microstructural studies, let alone for a study of the electrical properties.

In reference 370, a more thorough study was made of the sections $Ga_2Te_3 - A^2B^6$, where A^2 – Zn, Cd, Hg, B^6 – Te. Samples of the system $Ga_2Te_3 - A^2B^6$ were synthesized by fusion of the binary compounds, followed by prolonged annealing lasting 20 to 30 days at the appropriate temperatures. An X-ray diffraction study of the change in the lattice spacing with the composition showed that miscibility gaps (according to the author's observations, somewhat greater than in the corresponding systems with In_2Te_3) and regions of ordering are present in all the systems (Figs. 29-31).

In the system Ga_2Te_3 – ZnTe (Fig. 29), a two-phase region is observed from 55 to 73 mole % Ga_2Te_3. Most of the alloys show a normal zinc blende structure, and, in the vicinity of the composition with 75 mole %, there is a region where chalcopyrite exists. The exact region of this ordering has not been determined, but alloys with 73 mole % Ga_2Te_3 are ordered.

A miscibility gap is observed in the system Ga_2Te_3 – CdTe (Fig. 30) in the region of 42 to 73 mole % Ga_2Te_3. At about 75 mole % Ga_2Te_3, an ordered type of chalcopyrite appears. In addition to the main region of the gap at about 85 mole %, there appears a second region of two phases which are located between 83 and 87 mole % Ga_2Te_3. The regions of ordering are somewhat wider in this system than in Ga_2Te_3 – ZnTe.

In the system Ga_2Te_3 – HgTe (Fig. 31), as in the other two, the two-phase region is located between 42 and 73 mole % Ga_2Te_3. Normal chalcopyrite appears on the side of Ga_2Te_3 of this region.

Ordering is no longer observed in the composition with 85 mole % Ga_2Te_3. On the side of HgTe, a second ordered structure was observed in the region from the gap at 20 mole % Ga_2Te_3. The type of this structure was not determined.

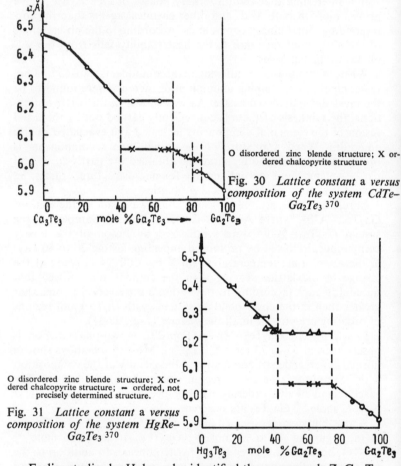

Fig. 30 *Lattice constant* a *versus composition of the system* $CdTe$–Ga_2Te_3 [370]

O disordered zinc blende structure; X ordered chalcopyrite structure

O disordered zinc blende structure; X ordered chalcopyrite structure; — ordered, not precisely determined structure.

Fig. 31 *Lattice constant* a *versus composition of the system* $HgRe$–Ga_2Te_3 [370]

Earlier studies by Hahn, who identified the compounds $ZnGa_2Te_4$ and $CdGa_2Te_4$ with the structure of chalcopyrite, do not contradict the data of reference 370. Both points corresponding to these compounds lie in the region of ordering on the diagram given in the latter work.

In the system Ga_2Te_3 – $ZnTe$ (see Fig. 29), the dependence of the forbidden gap width on the composition was determined for one-phase samples. The left- and right-hand portions of the curve (separated by the region of the gap) appear to be parts of a single, smooth dependence. The general slope remains unchanged in the region of the gap. Taking into account the difficulties involved in attaining an equilibrium, one can assume that in this system solid solutions exist over the entire range of compositions.

In reference 384, substitutional solid solutions were obtained in the system In_2Se_3 – $HgSe$ over the concentration range from $HgSe$ to $HgSe \cdot In_2Se_3$. The authors did not observe any ordering in the alloys of this system.

Studies made in reference 385 showed that the phases $ZnIn_2S_4$, $CdIn_2Se_4$, $HgIn_2Se_4$, $ZnIn_2Te_4$, $CdIn_2Te_4$ form via the peritectic reaction. Only $HgIn_2Te_4$ melts congruently. The common method of recrystallization was developed precisely for compounds which melt congruently. This method was modified for peritectic compositions[386].

The modified method was employed in the preparation and homogenization of samples of $CdIn_2Te_4$, whose electrical properties were studied[385]. Electrical properties of phases of this type were also studied in references 387, 388. All the substances were found to be typical semiconductors.

Table 37 shows the melting points and forbidden gap widths of compounds of type $A^2B_2^3C_4^6$ as given by the above-mentioned sources.

Table 37 Melting point and forbidden gap width of certain compounds of type $A^2B_2^3C_4^6$

Compound	T, °C	ΔE, eV
$ZnIn_2Se_4$	980 ± 10	2·6
$CdIn_2Se_4$	915 ± 5	1·3
$HgIn_2Se_4$	830 ± 10	0·6
$ZnIn_2Te_4$	800 ± 5	0·86—1·4
$CdIn_2Te_4$	785 ± 2	0·92—1·0
$HgIn_2Te_4$	708 ± 2	0·86—1·25

As a supplement to Table 37, other data for $CdIn_2Te_4$, which was apparently obtained in the purest form, should be cited.

At room temperature, samples of very pure $CdIn_2Te_4$ had electronic conduction, a carrier concentration $n = 10^{14}$ cm^{-3} and a mobility of 4000 cm^2/V sec[385]. In single crystals of $HgIn_2Te_4$ obtained by Bridgman's method, the electron mobility at $n = 3 \times 10^{15}$ cm^{-3}

was found to be 200 cm^2/V sec[388]. These are comparatively high values for compounds with structural vacancies. This obviously accounts for the interest shown in these compounds as materials for practical applications.

PHASES OF TYPE $A_2^2B^4C_4^6$

Recently, substances of this type were synthesized and studied from a structural point of view[35]. The synthesis was carried out by the usual method. Table 38 shows some properties of these substances.

Table 38   Structure, lattice spacings and density of substances of type $A_2^2B^4C_4^6$

Formula	Structure	Lattice spacings, Å			Specific gravity, g/cm^3	
		a	c	c/a	calc.	exp.
Zn$_2$GeS$_4$	Sphalerite	5·43$_6$	—	—	3·42$_7$	3·26$_0$
Zn$_2$GeSe$_4$	ditto	5·64$_6$	—	—	4·78$_9$	4·53$_2$
Cd$_2$GeS$_4$	Hexagonal-rhombohedral	7·1$_3$	35·1	4·9$_2$	4·11$_5$	3·82$_4$
Cd$_2$GeSe$_4$	ditto	7·4$_1$	36·2	4·8$_9$	5·44$_4$	5·19$_4$
Hg$_2$GeS$_4$	ditto	7·1$_7$	34·9	4·68$_6$	5·78$_9$	6·61$_1$
Hg$_2$GeSe$_4$	Thiogallate	5·69$_1$	11·28$_0$	1·98$_2$	7·17$_9$	7·09$_0$

The compounds Zn$_2$GeS$_4$ and Zn$_2$GeSe$_4$ have a wide region of homogeneity along the sections ZnS – GeS$_2$ and ZnSe – GeSe$_2$, and all the alloys crystallize in the sphalerite structure. The compounds Cd$_2$GeS$_4$, Cd$_2$GeSe$_4$ and Hg$_2$GeS$_4$ crystallize in the same structural type (hexagonal-rhombohedral), which has not, so far, been identified.

Hg$_2$GeSe$_4$ crystallizes in the same structural type as the majority of the above described compounds $A^2B_2^3C_4^6(S_4^2 - 14)$.

It is suggested that the compounds listed in Table 38 belong to the category of semiconductors.

PHASES OF TYPE $A_3^3B^5C_3^6$*

The substances $A_3^3B^5C_3^6$ were studied as alloys of the sections $A^3B^5 - A_2^3B_3^6$. Heterovalent substitution in these types of substance was first discovered in 1956[18]. To date, a large number of such systems have been studied (Table 39).

---

* No ordering was observed in the majority of the systems studied. Moreover, these systems are characterized by the formation of wide regions of homogeneity in the form of substitutional solid solutions (heterovalent substitution). All this suggests that substances of type $A_3^3B^5C_3^6$ are not compounds.

Table 39 Interaction of components in $A^3B^5 - A_2^3B_3^6$ systems

System	Type of interaction	Literature sources
AlSb—Al$_2$Te$_3$	Substitutional solid solutions in the concentration range of $\sim$ 45 mole % (from the AlSb end)	390
GaP—Ga$_2$S$_3$	Substitutional solid solutions in the concentration range of $\sim$ 50 mole % (from the GaP end)*	391
GaP—Ga$_2$Se$_3$	Substitutional solid solutions over the entire concentration range	391
GaP—Ga$_2$Te$_3$	Solid solutions in the concentration range of $\sim$ 50 mole % (from the Ga$_2$Te$_3$ end)	391
GaAs—Ga$_2$S$_3$	Substitutional solid solutions, $\sim$ 55 mole % (from the Ga$_2$S$_3$ end)	392
GaAs—Ga$_2$Se$_3$	Solid solutions over the entire concentration range	18, 377
GaAs—Ga$_2$Te$_3$	Substitutional solid solutions in the concentration range of $\sim$ 65 mole % (from the Ga$_2$Te$_3$ end)	377
GaSb—Ga$_2$Se$_3$	Complex interaction, details not studied	377
GaSb—Ga$_2$Te$_3$	ditto	393
InP—In$_2$Se$_3$	Solid solutions in the concentration range of $\sim$ 75 mole % (from the InP end)	279, 394
InAs—In$_2$Se$_3$	Solid solutions in the concentration range of $\sim$ 90 mole % (from the InAs end)	395, 396
InAs—In$_2$Te$_3$	Substitutional solid solutions over the entire concentration range	377, 397
InSb—In$_2$Te$_3$	Substitutional solid solutions, 15 mole % (from the InSb end)	398, 399

*Here and below all the mole per cent values have been calculated from the Sections 3 $A^3B^5 - A_2^3B_3^6$

The system AlSb – Al$_2$Te$_3$. In a study of the system AlSb – Al$_2$Te$_3$, all the alloys were prepared from the components or from the corresponding master alloys in an argon atmosphere, in corundum crucibles, and in resistance or induction furnaces[390, 400]. In reference 390, the authors established the general character of the fusibility diagram and discovered an appreciable region of solid solutions. A eutectic point was discovered at 85 mole % AlSb. The limits of solubility of Al$_2$Te$_4$ in AlSb were determined from a more detailed study of the region of the solid solution based on AlSb[400].

A study of the microstructure of slowly cooled alloys established that at $\sim$ 800° the solid solution in the system AlSb – Al$_2$Te$_3$ extends to $\sim$ 45 wt % Al$_2$Te$_3$. The samples containing more than 45 wt % Al$_2$Te$_3$ consist of two phases. X-ray diffraction photographs of these revealed a decrease in the lattice spacing of the solid solution of Al$_2$Te$_3$ in AlSb. The dependence is almost linear. In the two-phase alloys, a certain increase in the lattice constant was observed as compared to the alloy containing 48 wt % Al_2Te_4. The authors explain this by saying that equilibrium is much more difficult to attain in the two-phase region.

Measurements of the density of the alloys of the system AlSb – Al_2Te_3 in the region of the solid solution show that it increases with increasing concentration of Al_2Te_3.

The electrical properties of the alloys in the region of the solid solution based on AlSb were studied. The temperature dependence of the electrical conductivity in the system AlSb – Al_2Te_3 is of the same nature as that of compounds of type A^3B^5, with the exception of the portion of the curve in the region of the electrical conductivity jump, which corresponds to the transition from solid to liquid and to the presence of the two-phase region.

Measurement of the inverse voltage in alloys of the solid solution showed that as the Al_2Te_3 content rose, the inverse voltage increased from 3-5 V for AlSb to 18 V for the alloy with 12·68 wt % Al_2Te_3.

The systems GaP – Ga_2S_4; GaP – Ga_2Se_4; GaP – Ga_2Te_4. Alloys of these systems have been studied insufficiently. Tentative solubility limits are given in Table 39. The existence of solid solutions was established over the entire concentration range in the system GaP – Ga_2Se_3 [391]. The authors noted a partial ordering in certain alloys; the electrical properties have not been studied in any of the three systems.

The system $GaAs$ – Ga_2S_3 [392]. The alloys of this system were prepared by direct fusion of the components in evacuated and sealed quartz ampoules with vibratory stirring. Annealing at 900° lasted from 880 to 1440 hr. The addition of 3 to 4 per cent gallium arsenide led to the formation of a solid solution with a sphalerite structure whose lattice spacings changed linearly (increased) with the concentration of the arsenide up to 56 mole % Ga_2S_3.

The alloys containing 56 to 8 mole % Ga_2S_3 could not be homogenized, and the ingots had many phases. The formation of a one-phase region with a sphalerite structure was again observed in the vicinity of gallium arsenide.

Within the confines of the homogeneous region (100 — 56 mole % Ga_2S_3), the microhardness changed along a curve with a maximum at 43 mole % Ga_2S_3. $H_{max} \simeq$ 1070 kg/mm^2 (both here and later, the mole per cent values for the systems are calculated from the relation $3A^3B^5 \to A_2^3B_3^6$). Using the data of the thermal analysis, which showed a consistent pattern (typical of solid solutions) for the alloys over the entire concentration range, the authors suggest that when the alloys are homogenized more effectively, it will be possible to obtain a continuous series of substitutional solid solutions along this section. They report the alloys to be photosensitive in the visible portion of the spectrum. According to their observations, non-homogeneous samples displayed a higher sensitivity to light than did homogeneous ones.

The system $GaAs - Ga_2Se_3$. The alloys were synthesized by the usual method, i.e., fusing the elemental components in evacuated quartz ampoules. In reference 18, the alloys were not subjected to prolonged heating, and a proportion of them was not in a state of equilibrium; this corresponded to the region of inhomogeneity in the middle part of the section observed during the X-ray investigation. However, data obtained from study of the microstructure (the polished sections were etched with nitric acid) and from thermal analysis showed that the system represents a continuous series of solid solutions. The authors of a later work[377], who annealed the alloys of the system for several months, arrived at the same conclusion.

It was noted that the alloys close in composition to Ga_2Se_3 were homogenized much more rapidly than the alloys rich in gallium arsenide. The sphalerite structure was observed over the entire concentration range, and the lattice spacings departed somewhat from Vegard's law in the region close to Ga_2Se_3.

According to reference 18, the density calculated from X-ray analysis is in good agreement with the density determined experimentally for alloys close to gallium arsenide. Departures from the usual relationships observed in the properties of alloys rich in the defect component undoubtedly indicate the complex nature of the heterovalent substitution in defect structures. It is interesting to note that no one has observed any indications of ordering in the structure of the alloys of this system.

These alloys possess photosensitivity, with a maximum which shifts smoothly from 0·7 to 1·2μ (from Ga_2Se_3 to GaAs, respectively).

A study was made of the electrical properties of alloys of the system $GaAs - Ga_2Se_3$ [376, 401]. It was found that when small amounts of the selenide are introduced into gallium arsenide, the electrical conductivity increases. It reaches a maximum, after which it decreases monotonically down to values corresponding to the conductivity of gallium selenide (Fig. 32). The authors explain this phenomenon by saying that the initial portions of selenium act as the donor impurity at the lattice points of gallium arsenide; as is known from other studies, this increases the conductivity. The decrease in electrical conductivity is due to the formation of a poorly conducting solid solution of gallium selenide and gallium arsenide. The temperature dependence of the conductivity and the Hall constant were used to determine the forbidden gap width. The values were in good agreement at high temperatures.

Fig. 33 shows the dependence of the thermo-emf of the alloys on composition. The character of the change corresponds to the fact that gallium arsenide possesses electronic conduction, and gallium selenide, hole conduction.

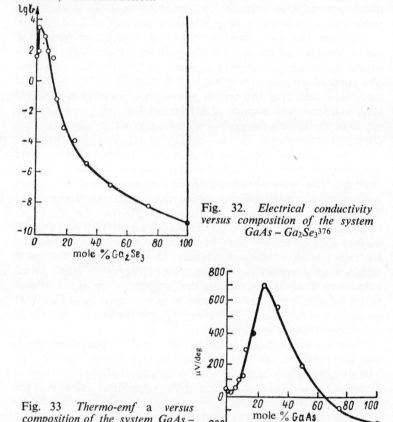

Fig. 32. *Electrical conductivity versus composition of the system* $GaAs - Ga_2Se_3$[376]

Fig. 33 *Thermo-emf a versus composition of the system* $GaAs - Ga_2Se_3$[401]

The system $GaAs - Ga_2Te_3$. A study of this section[377] showed that the alloys are homogenized only with great difficulty. Annealing at a gradually rising temperature was carried out over a period of two months. The region of substitutional solid solutions extends from 100 to 35 mole % GaTe. In alloys with a large content of gallium arsenide, two-phase regions were observed. The lattice spacings of the sphalerite structure in the region of the solid solutions changed along a curve located under Vegard's line, with a hump oriented toward the horizontal axis. The authors report that they were not able to observe any ordering in the alloys of this system.

The system $GaSb - Ga_2Se_3$ has been studied very little. Only in reference 377, where the system $GaSb - Ga_2Se_3$ of the composition 1:1 was synthesized, is it reported that the Debye powder patterns had a large number of lines. The same applies to the system $GaSb-Ga_2Te_3$.

The system $GaSb - Ga_2Te_3$ was studied in reference 393 as one of the sections of the ternary system $Ga - Sb - Te$. The samples were prepared by fusing the initial binary components in evacuated quartz ampoules. The cooling rate of the alloys in the crystallization range was 5°/hr. The authors found that in the section $GaSb - GaTe$ there is a region of a solid solution of GaTe in GaSb, which extends to 16.4 mole % GaTe, and, in the authors' words, "covers a part of the alloys of the section $GaSb - Ga_2Te_3$."

The systems $InP - In_2S_3$ and $InP - In_2Te_3$[279]. No solid solutions were observed in these systems. It is assumed that they can exist only in a narrow concentration range in the vicinity of the initial binary compounds.

The system $InP - In_2Se_3$. In reference 279 it was reported that this system contains an appreciable region of substitutional solid solutions. In the following work[394], seven alloys of this section were studied by fusion of the components in sealed and evacuated quartz ampoules with vibratory stirring. X-ray diffraction analysis and investigations of the microstructure and microhardness established that the alloys from 100 mole % InP to 75 mole % 3InP are single-phase alloys and have a sphalerite structure.

In reference 402, the existence of substitutional solid solutions was established in the region of concentrations from 100 to 14·1 mole % 3InP (in reference 402 the mole % values are calculated on the basis of InP, not 3InP, and for this reason the authors give 33 mole % InP). The lattice spacings of the alloys in the region of the solid solution decreased with an increase in the selenium content; the densities, determined pycnometrically and from X-ray data, were in good mutual agreement. The alloys were obtained and homogenized at 650°C.

Finally, in reference 391, the conditions of synthesis were made to conform to the data of the thermal analysis of the alloys, thus enabling the authors to obtain a region of substitutional solid solutions from 100 to 50 mole % indium phosphide.

Depending upon the concentration, the electrical conductivity of the alloys increased at first by more than one order of magnitude when the amount of selenide added was very small, then decreased gradually [403]. Such changes in the conductivity with the composition were also observed in other similar systems.

The system $InAs - In_2S_3$. Reference 404 offers evidence for the existence of a region of solid solutions up to 50 mole % InAs.

The system $InAs - In_2Se_3$. In the studies[395,396,404], the substances were synthesized in quartz beakers placed in sealed quartz ampoules filled with argon. The samples were annealed for 100 to 1100 hr at 600°C.

A study of the microstructure, X-ray phase analysis and thermal analysis showed that the system contains solid solutions with a sphalerite structure from 100 to 20 mole % indium arsenide. After prolonged annealing, the limit of formation of the solid solution shifts to ~ 10 mole % indium arsenide.

The alloys of the substances of this system were subjected to homogenization by annealing[405] in a reducing or inert gaseous medium at 600 to 700°C and 800 atm pressure for 250 hr. Following this treatment, the microhardness of the alloys shows much less scatter, and its absolute value increases. The microstructure of the alloys changes considerably; the dendrites disappear (the polished sections were etched with a mixture of 3% hydrogen peroxide with ammonia in the ratio of 1:3). The Debye powder patterns showed distinct lines corresponding to the sphalerite structure. No solid solution was observed in the vicinity of In_2Se_3.

Fig. 34 shows the phase diagram of the section $InAs \cdot In_2Se_3$ [396].* In the region of the single-phase alloys, the changes in the lattice spacing with the composition obey Vegard's law.

The microhardness has a maximum of ~ 440 kg/mm² at the composition $3InAs \cdot In_2Se_3$.

When alloys of the composition $InAs \cdot In_2Se_3$ were analysed by X-ray diffraction, lines were seen in addition to the fundamental ones, but could not be interpreted, since they belonged to a superlattice. Moreover, the general character of the changes in the properties of the system did not give any indication of ordering. X-ray diffraction analysis did not reveal any ordering in the solid solution of this system either.

The electrical properties of the alloys were investigated. The conductivity in the single-phase alloys increased by almost two orders of magnitude when only a small amount of indium selenide was added, and then gradually decreased. The explanation given for this

---

* When the conversion is made to 3InAs, the data of Fig. 34 correspond to those given in the text.

phenomenon is the same as that offered in the case of the system GaAs – Ga$_2$Se$_3$. The carrier mobility dropped sharply from indium arsenide to the compositions InAs·In$_2$Se$_3$ and 2InAs·3In$_2$Se$_3$. All the samples had electronic conduction.

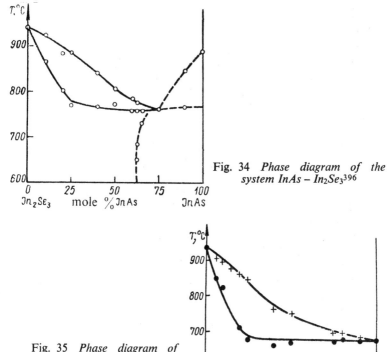

Fig. 34 *Phase diagram of the system InAs – In$_2$Se$_3$*[396]

Fig. 35 *Phase diagram of system InAs – In$_2$Te$_3$*[409]

In view of the fact that the majority of the samples had high carrier concentrations, it was not possible to reach the region of intrinsic conduction in measuring the temperature dependence of the electrical conductivity.

The forbidden gap width and its change with the composition were determined from the optical absorption edge[406]. It was found that the forbidden gap width of the alloys had values between 0·3 and 1·2 eV.

Reference 407 gives the results of an investigation of the thermal conductivity of the alloys of this system. From indium arsenide to the alloys containing indium selenide, the thermal conductivity decreases rapidly at first, then drops off in a nearly linear fashion.

The coefficient of linear expansion studied in this work increased with a rising concentration of indium selenide.

Reference 402 also noted the existence of solid solutions with a ZnS structure along the section InAs – InSe from 100 to 43 mole % InAs. The existence of a chemical compound of the composition In_2SeAs is postulated.

The system InAs – In_2Te_3. The alloys were synthesized under the same conditions as in the preceding system. In reference 408, the alloys were subjected to directed cooling. The existence of a continuous series of substitutional solid solutions was established over the entire concentration range. The lattice spacings change in an approximately linear fashion with composition[397], or follow a concave curve[377].

The phase diagram had the shape shown in Fig. 35. The microhardness of the alloys changed along a curve with a maximum at the composition $3InAs \cdot In_2Te_3$. Two etchants were used in succession to etch the polished sections: (1) a mixture of Perhydrol and acetic acid (1 : 1), and (2) a mixture of acetic and nitric acids (1 : 1).

No ordering was observed in any of the ternary alloys[397, 409]. The electrical and optical properties were studied; in this system, as in the preceding one, the electrical conductivity rose sharply when small amounts of tellurium were added to the arsenide, then decreased smoothly.

The data on the change in carrier mobility and forbidden gap width with composition differ somewhat in the reports of different authors. According to the data of reference 408, the forbidden gap width changes along a curve having a minimum, and according to reference 409, this change is smooth. The carrier mobility as given by 408 changes linearly with the composition at room temperature, and according to 409, it drops sharply from indium arsenide to $9InAs \cdot In_2Te_3$, then changes smoothly. The alloys had electronic conduction.

The authors of reference 408 point out the possibility of ordering in this system. In reference 410, this problem is discussed in more detail, and an analogy is drawn between alloys with defects in the lattice (indicated thus □ below), and nondefect substances with the structure of sphalerite.

Thus, it is thought that ordering is most probable in the following compositions:

In_3 □ $AsTe_3$ – 75 *mole* % In_2Te_3
In_5 □ As_3Te_3 – 50 *mole* % In_2Te_3
In_7 □ As_5Te_3 – 37·5 *mole* % In_2Te_3.

However, no experimental evidence has been produced thus far to indicate the existence of a superlattice in the ternary alloys of this system. In reference 402, a region of homogeneity was found along the section InAs – InTe from 100 to 50 mole % InAs.

The systems InSb – In_2S_3 and InSb – In_2Se_3. No solid solutions could be observed in these systems, according to reference 404.

The system InSb – In_2Te_3. In a study of the alloys of this section it was observed that solid solutions did not form in this system over a concentration range as wide as was the case earlier in the analogous systems GaAs – Ga_2Se_3 and InAs – In_2Te_3. It was found[399] that a chemical compound with the structure of NaCl was formed along another section, InSb – InTe. An X-ray diffraction study of the alloys of both sections was carried out [411].

It was found that even at the composition 9InSb·In_2Te_3, the Debye powder patterns show lines corresponding to an NaCl-type structure, in addition to the lines for the sphalerite structure. Then, as the composition changes further, the lines of the phase InTe appear, and in the vicinity of In_2Te_3, lines for the latter phase are observed.

The homogenization methods involving prolonged annealing and annealing under pressure did not produce any significant changes in the structure of the compositions studied[412]. On the basis of an X-ray study the authors suggest[411] that the formation of solid solutions over a narrow concentration range is possible in this section.

A study of the section InSb – InTe showed that all the alloys except the initial binary ones and the alloy InSb·3InTe, had at least two phases. The structure of the alloy InSb·InTe corresponds to the NaCl type with a lattice spacing $a = 6.128 \pm 0.003$ Å.

The density of the postulated compound In_3SbTe_4 was found to be 6.16 g/cm^3, and the line intensities of the Debye powder patterns of this substance were studied. Analysis of the data obtained made it possible to state that the authors of references 398, 411 were dealing with a new compound.

A more detailed study was made of the concentration range in the vicinity of indium antimonide in reference 399. The authors also observed that when the In_2Te_3 content exceeded 15 mole %, the Debye powder patterns showed the lines of a phase with the NaCl structure, but this was not studied further. They found that a solid solution of the telluride in InSb is formed up to 15 mole % In_2Te_3 (the alloys were annealed for 10 days at 500°C).

The authors prepared the alloys from the pure binary components: InSb had a carrier concentration of 10^{16} cm^{-3}, and In_2Te_3 10^{12} cm^{-3}. However, even when the amount of telluride added was small (of the order of 0·3 mole % In_2Te_3), the carrier concentration increased to $\sim 10^{19}$ cm^{-3}, then dropped off slowly. Thus, a phenomenon was also noted in this system which was similar to those observed in other systems of this type.

PHASES OF TYPE $A^2B^4C_3^6$

Substances of this type, which are analogues of the binary defect compounds $A_2^3B_3^6$, have been little studied. It is probable that in the systems $ZnS - GeS_2$ and $ZnSe - GeSe_2$, [35, 413], among the phases of variable composition with a tetrahedral structure there are phases corresponding to the formulae $ZnGeS_3$ and $ZnGeSe_3$, as well as the phases Zn_2GeS_4 and Zn_2GeSe_4, cited in Section C.

PHASES OF TYPE $A^3B^5C_4^6$

These substances are analogues of the binary compounds A^4B^6 The latter do not crystallize in structures with a tetrahedral coordination. Recently, a study was made of the ternary analogues that crystallize in structures analogous to SiS_2, $SiSe_2$ and $SiTe_2$, which serve as interesting examples of analogy in complex structural types[36]. As was stated above, in these structures the atoms are arranged tetrahedrally, but the tetrahedra form chains.

The six types of defect ternary phases enumerated exhaust the substances known up to the present time. As is evident from their description, substances having a tetrahedral structural configuration of the atoms are found in all the types. Quaternary and more complex heterovalent defect tetrahedral phases have been studied little[413], but the initial results are very interesting.

Part F
Excess Tetrahedral Phases

EXCESS tetrahedral phases, which may be regarded as resulting from the filling of tetrahedral or tetrahedral and octahedral vacancies formed by the close cubic or hexagonal packing of the nonmetal (see Section 3, Chapter 1), have been studied very little.

It may be assumed, however, that as far as their basic electronic properties are concerned, some of these substances are very promising, for example Cd_3As_2 (see Table 40) and others.

The accumulation of experimental data obtained by studying these substances in the pure state will make it possible to identify the principles governing the changes in these properties in groups of analogues and in isoelectronic series, as well as in families of sub-

Table 40 Some properties of compounds of type $A_3^I B^5$

Compound	Structure	Lattice spacing, Å	Sp. gravity, g/cm³	Method by which obtained	External appearance	Melting point, °C	Remarks	Literature source
Li₃N	Cubic	$a = 5.50$	1.3 (exp.)		Ruby-red crystals in the shape hexahedral of prisms		Heat of formation 47·166 kcal/mole	414–419
	Hexagonal	$a = 3.658$ $a = 3.882$	1.28 (X-ray)					
Li₃P	Hexagonal, type Na₃As	$a = 4.264$ $c = 7.579$	1.43	Combined fusion $t_{max} = 680°$	Reddish-brown mass, grey crystals under the microscope			418
Li₃As	Hexagonal, type Na₃As	$a = 4.387$ $c = 7.810$	2.42	Combined fusion, $t_{max} = 800°$	Brown, porous substance			418, 420
α-Li₃Sb (High temperature modification)	Hexagonal, type Na₃As	$a = 4.701$ $c = 8.309$	2.96	Chemical method	Bluish-grey alloy	1150–1300		418, 421
β-Li₃Sb	Cubic	$a = 6.559$	3.29	Chemical method		Temperature of α – β = transition, 650°C 1145		418, 421
Li₃Bi	Cubic	$a = 6.708$	5.03 (X-ray) 4.98 (exp.)	Fusion in argon atmosphere			Region homogeneity exists	422, 423
Na₃P	Hexagonal, type Na₃As	$a = 4.980$ $c = 8.797$	1.74	Chemical method and combined fusion	Brown amorphous mass (decomposes in air)		Heat of formation 32	418, 424, 425

Table 40 continued

Compound	Structure	Lattice spacing, Å	Sp. gravity, g/cm³	Method by which obtained	External appearance	Melting point, °C	Remarks	Literature source
Na₃As	Hexagonal, type Na₃As	$a = 5.088$ $c = 8.982$	2.36	Na vapour passed through heated As	Purple-brown substance		*kcal/mole*	418, 426, 430
Na₃Sb	Hexagonal, type Na₃As	$a = 5.355$ $c = 9.496$	2.67	Fusion in argon atmosphere	Blue-grey or purple-grey alloy with metallic lustre	856		418, 426, 427, 431, 432
Na₃Bi	Hexagonal, type Na₃As	$a = 5.448$ $c = 9.655$	3.70	Fusion in argon atmosphere	Ditto	775		418, 426, 427, 433
K₃P				Potassium ground in toluene heated with PCl₃, PCl₅	Black substance			418, 424, 434, 435
K₃As	Hexagonal, type Na₃As	$a = 5.782$ $c = 10.222$	2.14	Combined fusion $t_{max} = 800°$	Greenish alloy with metallic lustre			418, 429
K₃Sb	Hexagonal, type Na₃As	$a = 6.025$ $c = 10.693$	2.35		Substance with greenish - yellow lustre	812		418, 436
K₃Bi	Hexagonal, type Na₃As	$a = 6.178$ $c = 10.933$	2.98	Fusion in argon atmosphere	Greenish-yellow lustre	671		418, 437, 439

Compound	Structure	Lattice parameters		Preparation/Notes		Properties	References
Rb_3Sb	Hexagonal, type Na_3As	$a = 6.29$ $c = 11.17$					439
Rb_3Bi	Hexagonal, type Na_3As	$a = 6.42$ $c = 11.46$					439, 440
Cs_3Sb						Forbidden gap width, 1.3 eV	441-447
Cu_3N	Cubic	$a = 3.807$	5.84	Fusion of the elements Fusion of the elements Sublimation of the elements under vacuum Passing NH_3 through Cu_2O powder	640		175, 448
Cu_3P	Hexagonal	$a = 6.942$ $c = 7.098$ $a = 7.070$ $c = 7.135$		Fusion under vacuum	1023	Heat of formation, 32 kcal/mole	7, 440-451
Cu_3As	Hexagonal	$a = 7.088$ $c = 7.232$			830		7, 61
Cu_3Sb	Hexagonal	$a = 2.777$ $c = 4.367$					7, 61, 452
Ag_3N	Cubic	$a = 3.00$					7
AgP							453
$AgAs$				Structure not studied No compounds up to 20% Ag			454
Ag_3Sb	Hexagonal	$a = 2.984 \div 2.985$ $c = 4.803 \div 4.816$					7, 48, 61
AuP				Structure not studied			435

Table 41 Some properties

Compound	Structure	Lattice spacing, Å	Method of preparation	Specific gravity g/cm^3
α-Ag$_2$S	Cubic, 2 molecules in the cell	$a = 4.88$	Heating of the elements under vacuum at 350°C	
β-Ag$_2$S	Low-temperature modification, hexagonal			
α-Ag$_2$Se	Cubic, 2 molecules in the cell	$a = 4.983$	Fusion at 400°C	8·187
β-Ag$_2$Se	Tetragonal	$a = 7.06$ $c = 4.98$	Fusion under vacuum, zone refining	
	Orthorhombic	$a = 4.344$ $b = 7.111$ $c = 7.790$		
α-Ag$_2$Te	Cubic	$a = 6.572$	Fusion at temperature of red heat	8·318
β-Ag$_2$Te	Structure unknown			
α-Cu$_2$S	Antifluorite	$a = 5.564$		
α-Cu$_2$Se	Antifluorite	$a = 5.840$	Fusion of the elements under vacuum at 400°	
β-Cu$_2$Te	Hexagonal (stable at room temperature	$a = 4.23$ $c = 7.27$		

of $A_2^1B^6$ compounds

M. P., °C	ΔE, eV	μ_n, cm²/Vsec	μ_p, cm²/Vsec	Remarks	Literature sources
838					48, 455-462
$\beta \rightarrow \alpha$ at 179	1·3	63·5 (100°C)	19 (100°C)	Conductivity depends strongly on the sulphur content	458, 459, 461, 463, 465
897					48, 457, 458, 461, 466
	0·075	2000 (18—20°)	505 (18—20°)		458, 461, 464-467
958		15000 (at high temperature)			457, 461, 468
$\beta \rightarrow \alpha$ at 149·5		6000			461, 468, 469
1130 $\alpha \rightarrow \beta$ at 91·103					457
1113 $\beta \rightarrow$ at 110				Difficult to obtain in stoichiometric ratio	457, 458, 464, 470, 471
1111					61

Table 42 Some properties of compounds of type $A_2^2B^4$

Substance	Structure	Lattice spacing, Å	Method of preparation	Stability	M.P., °C	Heat of formation kcal/mole	ΔE, eV	μ_n cm²/V sec	μ_p cm²/V sec	Remarks	Literature source
Mg₂Si	Anti-fluorite	$a = 6.338$	Combined fusion in high-frequency furnaces in stoichiometric ratio		1102	—	0.77	406	56		16, 473, 476
Mg₂Ge	Anti-fluorite	$a = 6.380$	Ditto	Decomposes in air		—	0.77				16, 473, 474, 477
Mg₂Sn	Anti-fluorite	$a = 6.762$	Fusion in a crucible under a fusing agent at 1000° for 2.5 to 3 hr	Decomposes in air	795	5.7	0.32 0.36	3300	—	Magnetic susceptibility studied	16, 473, 474, 478-485
Mg₂Pb	Anti-fluorite	$a = 6.799$	Combined fusion	Ditto	550	—	—	—	—	Degenerated semiconductor $d = 2.12$ g/cm³	16, 473, 474, 482, 484, 486, 487-490
Ca₂Si	PbCl₂-type structure	$a = 9.002 \pm 0.016$ $b = 7.664 \pm 0.008$ $c = 4.799 \pm 0.006$	Reaction with peritectic transition		Temperature of peritectic transition, 910°		1.9	—			491-494

Substance	Structure	Lattice spacing, Å	Method of preparation	Stability	M.P. °C	Heat of formation, kcal/mole	ΔE, eV	μ_p cm²/V sec	μ_n cm²/V sec	Remarks	Literature source
Ca₂Ge	PbCl₂-type structure	$a = 9.069 \pm 0.009$ $b = 7.734 \pm 0.007$ $c = 4.834 \pm 0.004$	Ditto	Ditto	—	—	—	—			492
Ca₂Sn	PbCl₂-type structure	$a = 9.562 \pm 0.004$ $b = 7.975 \pm 0.004$ $c = 5.044 \pm 0.003$	Combined fusion	Ditto	1122	—	0.90	—			491, 495, 496
Ca₂Pb	PbCl₂-type structure	$a = 9.647 \pm 0.004$ $b = 8.072 \pm 0.004$ $c = 5.100 \pm 0.003$	Ditto	—	1110	—	0.46	—			491, 495, 497
Be₂C	Anti-fluorite	$a = 4.339$	Action of graphite on the metal during electrolysis							$d = 2.43$ g/cm³. Very hard, with abrasive properties similar to those of carborundum	498–504

Table 43 Some properties of

Compound	Structure	Lattice spacings, Å	Sp, gr., g/cm³	Method of preparation
Be_3N_2	Mn_2O_3-type	$a = 8.13$	2.70 (X-ray) 2.709 (exp.)	Nitrogenation (start at 900°C, end at 1100°C)
Be_3P_2	Mn_2O_3 type	$a = 10.15$	2.25 (X-ray) 2.234 (exp.)	Combined fusion at 750°C
Mg_3N_2	Mn_2O_3 type	$a = 9.9$	2.705 (X-ray) 2.712 (exp.)	Nitrogenation (start at 600–700°C, end at 866–1000°C)
Mg_3P_2	Mn_2O_3 type	$a = 12.03$ 12.01	2.058 (X-ray) 2.055 (exp.)	Phosphorus vapours in stream of H_2 passed through Mg powder
Mg_3As_2	Mn_2O_3 type	$a = 12.33$	3.138 (X-ray) 3.148 (exp.)	Arsenic vapours in stream of H_2 passed through Mg powder
Mg_3Sb_2	Li_2O_3-type	$a = 4.57$ $c = 7.22$		Combined fusion of components
Ca_3N_2	Cubic	$a = 11.40$–11.38	2.63 (X-ray) 2.64 (exp.)	Nitrogenation of calcium
Ca_3P_2	Cubic			
Sr_3N_2				
Sr_3P_2	Cubic			Synthesis from the elements
Mg_3Bi_2	Li_2O_3-type	$a = 4.64$ $c = 7.40$		Obtained like Mg_3P_2
Ba_3P_2				Reduction of barium phosphate by carbon black
Zn_3N_2	Cubic	$a = 9.74$	6.40 (X-ray) 6.222 (exp.)	Thermal decomposition of amide
Zn_3P_2	Tetragonal	$a = 8.09$ $c = 11.45$	4.54 (X-ray) 4.21–4.76 (exp.)	Fusion with subsequent sublimation
α-Zn_3As_2 2nd mod. exists	Tetragonal	$a = 8.31$ $c = 11.76$ $a = 11.78$ $c = 23.65$	5.66	Fusion and zone melting
Zn_3Sb_2		Metastable phase		
Cd_3N_2	Mn_2O_3-type	$a = 10.7$	7.67 (X-ray) 6.85 (exp.)	
Cd_3P_2	Tetragonal, Zn_3P_2-type	$a = 8.94$ $c = 12.28$	5.60 (X-ray) 5.95 (exp.)	Indirect fusion of the elements

compounds of type $A_3^2 B_2^5$

External appearance	Stability	M.P., °C	Heat of formation, kcal/mole	ΔE eV	μ cm²/V sec	Literature source
Greyish-white	Relatively stable		133·47			7, 505, 506
Chocolate-brown	Decomposes readily					7, 505, 536
Yellow	Hydrolyzes readily		116			505, 507-509
Yellow crystals	Decompose		128			7, 423, 451 505, 510-512, 536
Reddish-brown crystals						7, 423, 513, 514
Hexagonal platelets of metallic appearance ($H = 7$)		1228	0·65–0·8	0·65–0·7		423, 515-519
Dull black		1195				7, 505, 522, 523
Brown-red	Decomposes		120			451, 514, 536
Golden yellow						505
						505, 525, 536
		715				7, 423, 520, 521
Dirty black		3080				526, 527, 536
Black	Decomposes	226	5·3			7, 528-530
Grey colour metallic appearance			98·55	0·49 0·54 0·64		505, 531-536
Metallic grey ($H = 3$)		1115		0·93 1·00	10	505, 532, 537-542
		566				7
	Decomposes readily					528, 529
Grey metallic needles		700	24		Metal	449, 505, 532, 536, 543-545

Compound	Structure	Lattice spacings, Å	Sp. gr. g/cm³	Method of preparation
Cd_3As_2	Tetragonal Zn_3P_2-type	$a = 8.94$ $c = 12.65$	6.63 (X-ray) 6.25 (exp.)	Fusion of the elements in argon atmosphere
Cd_3Sb_2	Monoclinic $\beta = 100°14'$	$a = 7.20$ $b = 13.51$ $c = 6.16$		Metastable phase
Hg_3P_2	Tetragonal			
Ba_3N_2				

Table 44 Some properties of complex compounds

Compound	Structure	Lattice spacings, Å	Method of preparation
LiMgN	Antifluorite type	$a = 4.98$	Heating the binary compounds Li_3N and Mg_3N_2 together
LiMgP	Ditto	$a = 6.020$	Fusion (for homogenization heating at 650°)
LiMgAs	,, ,,	$a = 6.222$	
LiMgSb	,, ,,	$a = 6.623$	
LiMgBi	,, ,,	$a = 6.76$	
LiZnN	,, ,,	$a = 4.88$	Heating LiN and Zn_3N_2 together at 400° in a stream of NH_3
LiZnP	Antifluorite	$a = 5.78$	Fusion (for homogenization heating at 650°)
LiZnAs	Ditto	$a = 5.92$	Fusion (at 550°)

Table 43 cont.

External appearance	Stability	M.P., °C	Heat of formation, kcal/mole	ΔE eV	μ cm^2/V sec	Literature source
Grey, metallic colour (H < 3·5)		721		0·13–0·14	10·000 15·000 (in single crystal)	531, 532, 539, 546–548
		423				7, 549
						536
Dirty black						505

(analogues of type $A_2^2B^4$, $A_3^2B_2^5$ and $A_4^3B_3^4$)

External appearance	Stability	Sp. gr., g/cm^3	Remarks	Literature source
Reddish-brown	Hydrolyzes readily	2·41		30, 550
Brown, nonmetallic	Decomposes		Brittle	551, 552
			ΔE could not be measured	30, 553–555
			$\Delta E = 0{\cdot}7\ eV$	30, 554, 555
			$\Delta E = 0{\cdot}4\ eV$	30, 554, 555
Black	Hydrolyzes readily	4·61		30, 550
Brown, nonmetallic	Decomposes			551, 552
Ditto	Decomposes			551, 552

Compound	Structure	Lattice spacings, Å	Method of preparation
AgMgAs	Antifluorite	$a = 6\cdot25$	
AgZnAs	Ditto	$a = 5\cdot91$	Fusion of components
NaMgAs	,, ,,		
NaZnAs	,, ,,	$a = 5\cdot91$	Fusion of components
CuZnAs			
CuMgAs			
CuMgSb	,, ,,	$a = 6\cdot164$	
CuMgBi	,, ,,	$a = 6\cdot268$	
NiMgSb	,, ,,	$a = 6\cdot048$	Fusion of components ($T_{max} = 1400°$)
NiMgBi	,, ,,	$a = 6\cdot166$	Ditto
CuMnSb	,, ,,	$a = 6\cdot066$	Ditto
CoMnSb	,, ,,	$a = 5\cdot891$	Ditto
CuCdSb	,, ,,	$a = 6\cdot27$	Fusion of components at 800°
Ag$_3$SBr	Perovskite		
Ag$_3$SJ	Ditto		
Li$_3$AlN$_2$	Antifluorite	$a = 9\cdot48$	Li$_3$N+Al (730°) Li$_3$Al+N$_2$
Li$_3$AlP$_2$	Deformed Antifluorite Rhombic	$a = 11\cdot49$ $b = 11\cdot63$ $c = 11\cdot75$	Li$_3$Al+P (600°–700°) Li$_3$P+Al+P (650°–700°)
Li$_3$AlAs$_2$	Ditto	$a = 11\cdot88$ $b = 12\cdot00$ $c = 12\cdot13$	Li$_3$Al+As Li$_3$As+AlAs
Li$_3$GaN$_2$	Antifluorite	$a = 9\cdot61$	Li$_3$Ga+N$_2$ (600°)

Table 44 cont.

External appearance	Stability	Sp. gr., g/cm³	Remarks	Literature source
Bluish-grey, metallic			M. P., = 720° C	30, 556 557 558 558
Dark-grey, metallic	Stable		M. P., = 680° C	558 7
			Metal	559
Metallic character				30, 557, 558 30, 557, 558 556 556
Ditto				556
Ditto				556
Greyish-blue colour				30
Black colour	Decomposes in air			32
Black colour	Decomposes in air			32
White to light-grey	Hydrolyzes readily; stable to 1000° in nitrogen	2·33		30, 550, 560
Yellowish-brown	Hydrolyzes		Saltlike	561
Reddish-brown	Hydrolyzes		Negligible conductivity Saltlike	561
Light-grey	Hydrolyzes readily	3·35	Very hard	30, 550

Compound	Structure	Lattice spacings, Å	Method of preparation
Li_3GaP_2	Antifluorite	$a = 11.76$	$Li_3P + GaP$
Li_3GaAs_2	Tetragonal	$a = 11.96$ $b = 12.15$	$Li_3As + GaAs$
Li_3FeN_2			
Li_5SiN_3	Cubic	$a = 9.46$	Heating together of binary nitrides (850–1300°)
Li_5SiP_3	,,	$a = 5.85$	Heating together of binary phosphides (500–1000°)
Li_5SiAs_3	,,	$a = 6.05$	Heating together of binary arsenides (800–1000°)
Li_5GeN_3	,,	$a = 9.90$	Heating together of binary nitrides in nitrogen
Li_5GeP_3	,,	$a = 5.89$	Heating together of binary phosphides (750–1000°)
Li_5GeAs_3	,,	$a = 6.09$	Heating together of binary arsenides (700–800°)
Li_5TiN_3	,,	$a = 9.72$	Heating together of binary nitrides
Li_5TiP_3	,,	$a = 5.97$	Heating together of binary phosphides (800–900°)
Li_5TiAs_3	,,	$a = 6.15$	Heating together of binary arsenides (600–700°)

Table 44 cont.

External appearance	Stability	Sp. gr., g/cm³	Remarks	Literature source
				29
				29
			Complex $Li_3N \times FeN$	560, 562
Greyish-yellow	Stable above 1300°	2·21 (X-ray) 2·23 (exp.)		563
Brown	Hydrolyzes			29
Greyish-black to black	Hydrolyzes			29
Yellowish	Hydrolyzes	2·98 (X-ray) 2·95 (exp.)		31, 563
Brown	Decomposes, hydrolyzes			29, 31
Greyish-black	Hydrolyzes			29, 31
Yellow	Hydrolyzes	2·42 (X-ray) 2·30 (exp.)		31, 563
Brown	Hydrolyzes			29, 31
Black	Hydrolyzes			29, 31

stances in which the tetrahedral vacancies of the packing of the nonmetal atoms are filled to various degrees.

Of great interest is the problem of the possible formation of solid solutions by substances of this group, both between themselves and with compounds having an electron concentration of four electrons per atom and a zinc blende structure. The formation of continuous substitutional solid solutions was observed in a group of magnesium compounds of the Mg_2X, where X is silicon, germanium, tin, or lead, and also in other types of excess phase.

The formation of solid solutions between excess phases of different types and compounds with the ZnS structure, i.e., the formation of phases of variable composition based on heterovalent substitution, would constitute evidence of the relationship between these types of structures.

It is possible that when such substitutional solid solutions are formed, the same phenomenon occurs as in phases of variable composition based on substances with the structures of chalcopyrite and zinc blende: the phenomenon of disordering. The preparation of these complex alloys is also unquestionably interesting from the practical point of view (primarily because of their potential use in the field of thermoelectricity).

The properties of some compounds of types $A_3^1B^5$, $A_2^1B^6$, $A_3^2B_2^5$ and $A_4^3B_3^4$ are given below in Tables 40-44.

CHAPTER 3

PRINCIPLES GOVERNING VARIATIONS IN THE PROPERTIES OF DIAMOND-LIKE SEMICONDUCTORS

1. Variation of Properties in Groups of Analogues

As the atomic number of the constituent elements increases in a group of diamond-like, binary and more complex phase semiconductors, all the properties change in a definite fashion which reflects the weakening of the bonds between the valence electrons and the atomic nuclei. This change, which is non-monotonic, results from a non-monotonic change in the electronic characteristics of the free atoms constituting the compounds.

D. I. Mendeleev surmised that the variation of the properties of the elements in the vertical columns of the periodic system may be non-monotonic, and for this reason, in predicting the properties of any element that was still unknown, he averaged the values not only vertically but also horizontally[564].

The phenomenon of the non-monotonic variation of properties was noted by Byron in similar chemical compounds of elements of the main subgroups of groups V, VI, and VII[565]. He termed this phenomenon "secondary periodicity".

S. A. Shchukarev showed that the secondary periodicity of the properties of compounds is a direct reflection of the periodic trend of the ionization energies of the atoms of elements in the main subgroups in the direction from the top to the bottom of the system[564].

The periodic trend of the energy characteristics of isolated atoms and ions of the elements in the main subgroups is due to the filling of the d and f levels of the electron shells, which leads to a rein-

194   CHEMISTRY OF DIAMOND-LIKE SEMICONDUCTORS

forcement of the bond between the nucleus and the outer *s* and *p* electrons.

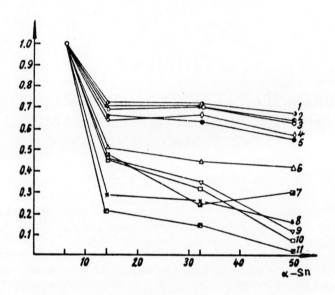

1, electron affinity constant; 2, first ionization potential; 3, reciprocal of lattice constant; 4, boiling point; 5, total ionization potential; 6, heat of sublimation; 7, reciprocal of electron mobility; 8, reciprocal of gram-atomic volume; 9, dielectric constant; 10, melting point; 11, forbidden gap width.

Fig. 36 *Properties of substances in the diamond group versus the atomic number of the element* n *(properties of diamond taken as unity)*[200]

It is necessary to emphasize the complexity of the change taking place in the properties of compounds of the elements in the main subgroups. By overlooking this fact, certain investigators have admitted an over-simplification of the periodic law.

In accordance with the above, the total ionization potentials of the atoms of group IV elements show a very definite stepwise change with increasing atomic number[564].

Inasmuch as the rough schematic model of substances of the diamond group contains ionic cores in place of elements, i.e., atoms stripped of their valence electrons and having completely filled outermost shells, the principal energy characteristics of the atoms are the electron affinity constants (and also group ionization potentials). Fig. 36 shows the variation in the properties of the substances of the diamond group with the atomic number (for convenience, the properties of diamond are taken as unity, and reciprocal values are taken for certain quantities). It is apparent from the figure that the stepwise character of the variation is particularly obvious in the energy characteristics of the free atoms.

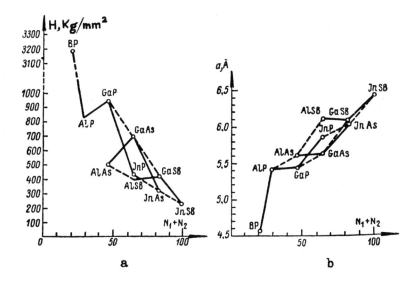

Fig. 37 *Hardness (a) and lattice spacing (b) versus the mean atomic number of A^3B^5 compounds*

In a smoother form, caused by the complication of the functional dependence, the secondary periodicity, which is expressed in the ordering of the angles of ascent and descent along the curve of property versus atomic number, shows up also in such parameters as the lattice constant, the gram-atomic volume in the solid state, the melting point, the dielectric constant, the sublimation energy, and the forbidden gap width.

The interatomic distances in a lattice of a single type, both for elements and for compounds, increase with the increasing atomic weight of the structural units constituting the lattice, as a direct result of the weakening of the atomic interaction caused by the steadily increasing screening of the atomic nuclei by new electron shells (Table 45). This is also associated with an increase in the gram-atomic volume. The hardness in the series of analogue elements and analogue compounds decreases with increasing atomic number, proving that the decrease in mechanical strength and the weakening of the cohesive forces are due to the same aforementioned reasons. The coefficients of thermal expansion increase correspondingly. Fig. 37 shows the change in microhardness and lattice spacings of substances of type A^3B^5.

The classical series of substances and compounds compiled by Goldschmidt[3] first demonstrated that, as the interatomic distances within one type of structure increase, the hardness decreases. As is evident from Fig. 37, the changes in the microhardness and interatomic distance with mean atomic number are of opposite sense. Attention is drawn to the abnormally low values of the microhardness of aluminium compounds, apparently due to the lattice spacings, which are close in magnitude to the lattice spacings of gallium compounds; this indicates a decrease in the strength of such a lattice (see Section 3 for the correlation between hardness and other quantities).

The melting points of the substances decrease as the atomic number increases. The melting point, like the hardness, depends on the cohesive forces and determines the limiting temperature to which a solid can be heated and still retain its shape. However, the rupture of the bonds which takes place during the transition from the solid to the liquid state is not always of the same type. In the description of silicon and germanium given above, it was stated that when these substances melt, the homopolar bonds are transformed into metallic-type bonds, but fusion is not characterized by such a process in all instances involving diamond-like semiconductors.

There exist[63] three principal types of change in electronic conduction (and hence, changes in bond character) in going from the solid to the liquid state. The first two types pertain to solids having semiconducting properties, while the third pertains to metals. The group

Table 45 Variation of physicochemical and physical properties in series of analogues

Substance	Lattice spacing, Å	Microhardness, kg/mm²	ΔE, eV	μ_n, cm²/V sec	μ_p, cm²/V sec	Sp. gr., g/cm³	Heat of sublimation, kcal	Coefficient of linear expansion, $\alpha \times 10^6$ deg⁻¹	Refractive index	Dielectric constant, ε	Melting point °C	Thermal conductivity, cal/cm sec deg;	Debye temperature, °K
C (diamond)	3.5597	8820 ± 1380	5.6	1800	1200	3.51	171.7	1.18	2.42	5.7	—	0.20	658 ± 6
Si	5.4198	1150 ± 110	1.21	1550	—	2.328	88.04	4.2	—	12	1420	0.14	362 ± 6
Ge	5.647	780 ± 79	0.78	4400	—	5.323	78.44	6.1	—	16	936	при 300° C	230
α-Sn	6.4912	—	0.08	3000	—	5.765	—	—	—	—	—	0.036	—
BN$_w$	{$a = 2.504$ $c = 6.661$}	—	4.6—3.6	—	—	2.25	—	10.5	—	4.15	—	—	598
BN$_s$	3.615	3200	—	—	—	3.45	—	7.5	3.0	—	—	—	—
BP	4.537	—	5.9	—	—	2.89	—	—	—	—	—	—	—
BAs	4.777	—	—	—	—	—	—	—	—	—	—	—	—
AlN$_w$	{$a = 3.104$ $c = 3.965$}	—	3.8	—	—	3.26	—	—	—	—	2200	—	—
AlP	5.45	—	3.0	—	—	2.77	—	3.5	—	—	>1600	—	400
AlAs	5.6622	500 ± 20	2.16	—	400	3.79	—	—	3.0	—	1060	—	350
AlSb	6.1355	400 ± 20	1.6	200	—	4.15	—	—	—	—	1500	—	—
GaN$_w$	{$a = 3.180$ $c = 5.160$}	—	3.25	—	—	6.10	—	—	—	—	—	—	—
GaP	4.4505	940 ± 35	2.4	100	20	4.14	—	3.5	2.9	—	1350	0.125	303
GaAs	5.6534	700 ± 50	1.53	6000	400	5.4	—	5.8	3.2	11.1	1237	0.105	270
GaSb	6.0954	420	0.80	4000	700	5.65	—	6.9	3.7	—	703	—	—
InN$_w$	{$a = 3.53$ $c = 5.69$}	—	2.4	—	—	6.88	—	—	—	—	—	—	—
InP	5.8687	435 ± 20	1.34	3400	50	4.74	—	—	3.0	—	1070	—	242
InAs	6.0584	330 ± 10	0.45	23000	240	5.68	—	5.3	3.2	—	942	—	228
InSb	6.47877	220 ± 10	0.25	65000	1000	5.78	—	5.5	4.1	—	536	0.037	—

of diamond-like semiconductors which we are considering includes cases of both types of change associated with fusion. The first type includes certain compounds of type A^2B^6 (HgSe, HgTe), which retain their semiconducting properties on melting, whilst the second type includes silicon, germanium, and also compounds of type A^3B^5 – InSb, GaSb, etc., whose structure becomes rearranged during melting in the direction of an increased coordination number.

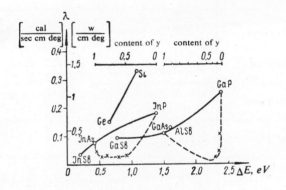

O elements or compounds; X solid solutions.

Fig. 38 *Thermal conductivity λ versus forbidden gap width ΔE for Si, Ge, compounds of type A^3B^5, and solid solutions InP – InAs and GaP – GaAs*[566]

The experimental data are still not sufficient to make it possible to follow the change in pattern of transformation of the bonds during melting in the various series of analogues, but references 63, 165, 217 make it possible to assume that the change taking place in the bond during melting is dependent upon the atomic numbers of the elements included in the composition of the substance.

From the data available in the literature it is evident that the Debye temperatures characterizing the degree of excitation of the various lattice vibrations decrease with increasing atomic number.

As was shown by studies of series of analogues (elements and binary compounds of type A^3B^5), thermal conductivity decreases with increasing atomic number [225, 566], for example in the series $InP - InAs - InSb$[567]; this accords with the theory of thermal conductivity (Fig. 38).

It is interesting to note that if binary compounds of type A^3B^5 having approximately equal molecular weights are compared, for example, AlSb, GaAs, and InP, the minimum thermal conductivity is observed in gallium arsenide, i.e., in the compounds where the atomic masses are the closest.

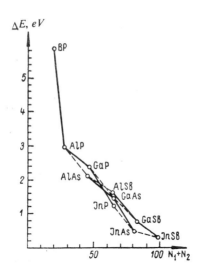

Fig. 39 *Forbidden gap width* ΔE *of* A^3B^5—*type compounds versus* **mean** *atomic number*

As can be readily observed, the external appearance of the substances changes with increasing atomic number, i.e., they assume a more metallic appearance. The "metallic" component of the bond, produced by the increasing screening of the nucleus, becomes superimposed on the principal covalent bond type in these substances.

The "metallization" of the bond in the series of analogues of elements and compounds causes a narrowing of the forbidden gap. This process may be visualized as follows: the weakening of the bond of electrons occurs inside the covalent "bridges"; the latter become more or less diffuse, so to speak, depending upon the atomic number. Owing to the weakening of the bond, the activation energy of the valence electrons (the forbidden gap width) decreases, but does not become equal to zero, as would be the case if some of the electrons migrated into the electron gas. The latter process is associated with an increase of the atomic number in such groups of analogues, and causes an abrupt jump in properties. The substance ceases to crystallize in a structure characterized by semiconducting properties, since the bonds cease to be directional. In the group of elemental semiconductors, such a jump occurs when grey tin is converted into white tin. The next element (lead) is a still more typical metal than white tin.

Fig. 39 shows the change in the forbidden gap width with the atomic number for compounds of type A^3B^5, from which can be seen that the change in ΔE is usually stepwise in character.

The dielectric constant is related to the electronic polarization, which depends in turn on the strength and nature of the bond between the electron and the nucleus. For this reason, the course of the variation of dielectric constant with atomic number is approximately the same as that of other quantities which are dependent on the same parameter.

The optical properties also change in accordance with the enhancement of metallic character and weakening of the covalent fraction of the bond: diamond absorbs light in the ultraviolet portion of the spectrum, and this explains its transparency and colourlessness. From silicon to germanium, the absorption band shifts from the visible to longer wavelengths.

The sensitivity maximum of the spectral distribution of the photoelectric effect occurs at wavelengths of incident light corresponding to the absorption band in the optical spectrum, and as a result it shifts in the same fashion.

In series of analogues, the index of refraction increases from the top to the bottom, since the refraction is caused by deformation of the outer portion of the electron cloud of the compound by the incident light, and weakening of the bond between the atoms naturally accompanies the deformation.

The changes in carrier mobility throughout a group of elemental semiconductors is complex and measurements are sometimes very difficult to perform. From the theory of the electrical conductivity of crystals it follows that as the atomic mass increases, other things being equal, the carrier mobility should rise. This takes place chiefly because of the smoothing out of the potential field of the crystal by the electron clouds of heavy atoms, and usually, as the atomic weight (atomic number) increases, the carrier mobility rises in groups of crystallochemical analogues. The high carrier mobility in substances of the diamond group is due to the special characteristics of the bond type. The principal characteristic consists in the presence of tetrahedrally oriented electronic "bridges", and leads to two important consequences:

(1) The amplitudes of the thermal vibrations of the atoms become very small (there exists a rigid framework of tetrahedral bonds);

(2) The effective mass of the conduction electrons is also small (the electron density does not drop to zero along the tetrahedral directions).

These features determine the high carrier mobility in diamond, silicon, germanium, grey tin, and in their binary analogues. The electrical conductivity of the substances in series of analogues increases with the atomic weight, whilst the thermoelectromotive force decreases.

The relationships governing the variation of properties in series of analogues of binary compounds and the elements – diamond, silicon, germanium, and tin – are generally the same, since they are caused by the very same process of metallization of the bond, although in the first case the metallization is superimposed on a mixed ionic-covalent bond, and in the second case, on a covalent bond. By analysing the vertical series of analogues which are binary compounds of types A^3B^5, A^2B^6, and A^1B^7, one can observe that the heats of formation diminish with increasing atomic weight, since the metallization of the elements constituting these compounds leads to a lesser change in the energy of the system as a result of the reaction. In these series (and in these series only, not in isoelectronic or any other series), the heat of formation may be used as an index of the stability of the chemical compound, since such series have one and the same fundamental type of bond on which the metallization becomes superimposed. Secondary periodicity manifests itself particularly often in the variation of this quantity. All of the above applies also to multicomponent phases of diamond-like structure.

In reference 568 it was shown that the energy of atomic dissociation seems to provide the most correct characterization of the bond energy in substances of types A^3B^5 and A^2B^6. This energy is calculated from the equation

$$\Omega = Q_p + S_m + \frac{1}{2}D_x,$$

where Q_p is the heat of formation from the elements in their standard states, S_m is the heat of sublimation of the metal, and D_x is the dissociation energy of a mole of the nonmetal in the gaseous state.

The atomic dissociation energies of substances of type A^2B^6 in series of analogues always decrease with increasing atomic number. For compounds of type A^3B^5, the energies of atomic dissociation obtained by interpolation decrease in the same direction. However, certain types of this group have their own peculiar characteristics in the variation of their properties.

It is interesting to examine the change in shape of the phase diagrams of compounds of type A^3B^5 with increasing atomic weight of one of the components (see Figs. 13-19). It is evident that as the melting points of the compounds decrease, the degree of degeneration of the eutectics formed by the metallic components and these compounds diminishes. The eutectic point is closest to the singular point in the heaviest compound, indium antimonide.

In reference 569 a suggestion is made that the ability of binary compounds to exist over a specific range of compositions is also subject to regular change in the periodic system.

In the case of compounds of type A^3B^5, attempts have been made to determine the region of homogeneity[109, 110, 570], but they ended in failure, apparently because the width of the homogeneous region of these compounds does not exceed the limits of error of the X-ray method. Nor were other methods successful in detecting the existence of a region of variable composition in the vicinity of the stoichiometric point[112, 571].

In the case of ternary compounds of the diamond-like group, the problem of the width of the region of homogeneity has its peculiar aspect (see Parts D and E). The formation and extent of regions of homogeneity with defect and excess tetrahedral phases and phases with a variable valence in ternary systems are unquestionably directly dependent on the position of the atoms in the periodic systems.

In quaternary and more complex systems where wide regions of homogeneity of nondefect tetrahedral phases are the rule (see Section 2, Part D), and where, in addition, regions of homogeneity of another type are possible, the above-mentioned relationships assume a still more complex form.

Semiconductors of the diamond-like group have been studied from the standpoint of their catalytic properties[572]; a definite relationship was observed between these and their electronic properties. It is very interesting to note that in series of analogues (for example, in the series ZnS – ZnSe – ZnTe; Ga_2Se_3 – Ga_2Te_3, etc.), the catalytic activity increases considerably in the region of intrinsic conduction[573, 574], i.e., that a regular variation with intrinsic conductivity was observed.

S. Z. Roginskii also notes that all diamond-like semiconductors have the same "catalytic spectrum", despite substantial quantitative differences in their activity[575]. This conclusion is important support for the idea that the group of diamond-like semiconductors (and possibly the whole family of substances with tetrahedral bonds, including defect and excess phases) is endowed with far-reaching similarities in properties that extend to many substances. An example of this similarity is to be found in study of the kinetics of solution of compounds of type A^3B^5, mentioned above, and in numerous studies of other properties.

As follows from Part C, there exists a large number of diamond-like phases which are formed by isovalent substitution. In analysing the properties of isovalent solid solutions, it should be noted that the properties of binary compounds in the vertical series, with the substitution of one of the atoms, change discontinuously with the substitution and obey the relationships described above.

Isovalent solid solutions form a transition between binary compounds. Their properties change continuously, for the most part filling smoothly the entire interval between the properties of the initial compounds. Experiments have shown that the formation of isovalent solid solutions between diamond-like compounds is a very common phenomenon. Previously unknown solid solutions (for instance, InP – GaP or InAs – InSb) have now been obtained, and, conversely, analogous systems are unknown in which the absence of substitutional solid solutions could be demonstrated for the entire concentration range. It is true that the process of formation of solid solutions by substances of this group is difficult, as we have seen from their description (see Part C), since the covalent bonds, being rigid, short-range bonds, prevent diffusion.

A continuous and smooth variation of properties makes it possible to select substances with predetermined properties. However, of no less interest are those instances of isovalent solid solutions in which the changes in properties are not smooth or are far from linear.

In these cases (which have not been thoroughly analysed, with the exception of the few mentioned in Part D), we are able to penetrate more deeply into the nature of the initial substances and its relation to the property being studied. For example, the characteristic variation of the forbidden gap width in the solid solution of silicon and germanium experimentally confirmed Herman's hypothesis, concerning the diverse structure of the energy bands of these elements, and helped to construct a more detailed representation of the pattern of the change in the valence band and conduction band with increasing silicon concentration.

The non-linear course of the variation in electrical properties is sometimes interesting from a practical standpoint. Thus, in the above-mentioned case, for example, this makes it possible to prepare alloys (in the region close to germanium) with a relatively small reduction in mobility and a substantial increase in the forbidden bandwidth as compared with germanium.

Of great interest in this respect are isovalent solid solutions of mercury chalcogenides, where the researchers first encountered a whole series of anomalies in electrical properties, the causes of which have not yet been elucidated.

A very interesting and, apparently, often useful hypothesis is that advanced by Regel concerning the stabilization of the lattice of the initial compounds in solid solutions, and the healing of defects, which play a significant part in the scattering of electrons.

According to a hypothesis first put forward by Ioffe[117, 576], in solid solutions the thermal conductivity of the substances changes by going through a minimum. This is shown in Fig. 39 for compounds of type A^3B^5 [566], and is a property which proves valuable from the standpoint of the use of solid solutions in thermoelectric devices.

2. Variation of Properties in Isoelectronic Series

Table 1 (see p. 3) listed the isoelectronic series of the crystallochemical group of diamond-like semiconductors. Also discussed in that section was the non-ionic type of interaction in substances of this group, first pointed out by Goldschmidt[3].

At the present time, the problem of the existence of covalent and ionic bonds in binary and more complex compounds of the diamond-like group is being investigated by various methods[577-580]; the model of the bond in compounds of type A^3B^5 is being actively debated.

The subject of discussion is the arrangement of the electrons in the covalent bridges between the ionic cores occupying the sites of the crystal lattice, and the effective charges on the atoms. There is no consensus of opinion on this subject[290, 593].

Let us consider the variation of the properties of substances in isoelectronic series, assuming that the ionic character, appearing in compounds of type A^3B^5, is enhanced in the direction of compounds of type A^2B^6 and even more in the direction of compounds of type A^1B^7 (Table 46).

Table 46 Some properties of diamond-like substances in the isoelectronic series

Substance	Mol weight	Lattice spacing, Å	Micro-hardness, kg/mm^2	ΔE, eV	μ_n	Brittleness criterion	M.P., °C	Dissociation kcal/mole [182]
Ge	145.20	5.647	780 ± 79	0.78	4400	4.4	936	89
GaAs	144.63	5.6534	700 ± 50	1.53	6000	3.0	1237	76?
ZnSe	144.34	5.653	135 ± 5	2.58—2.66	—	—	<1600	113.6
CuBr	143.46	5.811	—	2.94	—	—	504	—
Ge + α-Sn								
GaSb	191.48	6.0954	420	0.80	4000	1.8	703	69
ZnTe	192.99	6.087	90 ± 5	2.15	100	—	1239	—
CuJ	190.46	6.043	—	—	—	—	605	—
α-Sn	237.4	6.4912	—	0.08	3000	—	—	(70)
InSb	236.52	6.47877	220 ± 10	0.25	65000	2.2	536	121
CdTe	240.02	6.46	60 ± 5	1.41—1.47	600	0.6	1045	98.9
AgJ	234.80	6.47	—	2.8	30	—	—	—

It might be asked, first of all, why are the lattice spacings so similar in the isoelectronic series of diamond-like substances?

In compounds of the ionic series, we have an entirely different pattern of change of the lattice constants with changing valence of the elements. Thus, for example, NaCl and MgS, also formed by elements equidistant from group IV, like A^1B^7 and A^2B^6, have unit cell spacings which differ by approximately 10%, and the latter decrease when a cation of higher valence is involved. This tendency is very typical of ionic compounds, and is explained by the fact that atoms with a higher valence have a greater polarizing effect. An ionic bond causes the electron cloud to be displaced toward the anion and the electron density between the atoms approaches zero.

In compounds of the diamond-like group, for instance, CuBr and ZnSe, such a process does not seem to be decisive.

If one proceeds from the assumption that, in this group of substances, the basic type of interaction involves ionic cores joined by paired-electron tetrahedral bonds, and that all the valence electrons of the two atoms comprising the compound are redistributed regardless of which atom they belong to, it will appear natural that this bonding mechanism, common to all the compounds of the group, will give rise to equal interatomic distances if the characteristics of the cores are similar (as they are in isoelectronic series).

It is true that there is a certain increase in the interatomic distance in going from A^3B^5 to A^1B^7; this corresponds to the enhancement of ionic character as a result of an increase in the difference between the chemical nature of the constituting atoms. In addition, data on the cleavage of all binary compounds of this type[50], investigations of the energy of formation of Frenkel defects[580], and certain other data indicate a considerable proportion of ionic bond character.

The variation of the physicochemical and physical properties in isoelectronic series also indicates an increase in the ionic type of interaction from elements of group IV to compounds of type A^1B^7, in accordance with the increase of the difference in the chemical nature of the atoms constituting the compounds.

As is evident from Table 46, the microhardness drops and the energy of atomic dissociation also decreases in isoelectronic series when the lattice spacings and molecular weights and hence the molar volumes are constant. The melting point changes in a complex fashion.

The decrease in hardness established by Goldschmidt for such series has recently been confirmed by a series of new data[204]. A very characteristic feature is revealed here in the pattern of change of this quantity and also of the melting point.

Both the hardness and the melting point characterize the cohesive forces. However, as was stated in the preceding section, the melting point is related to these forces in a more complex fashion, and this is manifested clearly in diamond-like substances. It will suffice to cite as an example two substances of an isoelectronic series, zinc selenide and gallium arsenide. The melting points of these substances are about the same, but the microhardness of zinc selenide is at least five times smaller than that of gallium arsenide. It may be assumed that microhardness characterizes the stability of covalent bonds, and therefore decreases in isoelectronic series. The complex nature of the change of the melting point in isoelectronic series is due to the fact that when the liquid state is attained, not all the bonds are broken to the same extent in the different substances of this series. Apparently, the energy of atomic dissociation of substance in isoelectronic series also characterizes the strength of covalent bonds.

Lately, very interesting investigations of the properties of substances in isoelectronic series have been undertaken. Thus, for example, in reference 56, where a study was made of the physicomechanical properties of diamond-like substances, it was observed that as the ionic component of the bond increases, the breaking strength, the surface energy, the criterion of brittleness and the microhardness decrease. According to data given by the author of reference 56, germanium and the arsenides and antimonides of indium and gallium are characterized by high brittleness, whereas plasticity is clearly manifested in zinc selenide and cadmium selenide and telluride.

In references 75, 184, 250 it was established that the coefficients of thermal expansion of silicon, germanium, and grey tin become negative in the region of low temperatures. Substances of the isoelectronic series of germanium and grey tin were investigated in order to determine the influence of the ionic component of the bond on the anomalous course of the temperature dependence of the coefficient of expansion α. It was found that the temperature at which α shifts into the region of negative values rises, and the minimum of α becomes deeper, as the ionic character is reinforced. In reference 574 it was shown that as the ionic character of the bond increases, the catalytic activity decreases, and a tendency toward a rise in the activation energy of the catalytic reaction is observed.

As is evident from Table 46, the forbidden gap width in isoelectronic series increases with the ionic character. This is due to the fact that the difference between the maximum and minimum of the potential in the periodic field of the crystal increases, and the forbidden gap width is proportional to this difference[581]. The change of potential in the periodic field of a crystal with a covalent bond is less than the change of potential in an ionic crystal, for in the latter the electron density between the atoms drops to very low values, whilst in a crystal with covalent bonds the presence of covalent bridges levels the difference between the maximum and minimum of the potential.

In the view of Welker[581], the high carrier mobilities in compounds of type A^3B^5, as compared to A^2B^6, are accounted for precisely by the fact that the ionic character of the latter compounds is greater, and that the scattering of electrons by optically-induced vibrations of the lattice, which are enhanced along with the ionic character, become the chief factor determining the carrier mobility.

In compounds of type A^3B^5, the carrier mobilities usually exceed the mobilities of the elemental analogues. In reference 581, this fact is explained by the resonance stabilization of the bond in substances of type A^3B^5, and by the associated decrease in the amplitude of thermal vibrations of the lattice, which therefore scatter the charge carriers to a lesser extent. The latter factor accounts for the high carrier mobility in A^3B^5, but the scattering of electrons by the optical oscillations which arise together with the ionic bond plays a secondary part. In passing from compounds of type A^3B^5 to those of type A^2B^6, the most important part is played by the scattering by optical vibrations.

The qualitative picture provided by Welker repeats itself in different variants in later works as well. Without becoming involved in a criticism of the resonance theory used by Welker and the majority of foreign authors, it should be noted that the bond stabilization upon the addition of a small part of an ionic bond to the basic covalent bond may also be demonstrated without resorting to the resonance theory, as was done, for example, by Gubanov[582]. This qualitative picture is valid for isoelectronic series only, so that it may not be concluded that a correlation exists between the degree of ionic character and the forbidden gap width in all diamond-like semiconductors. The resonance picture does not account for the fact that the relative carrier mobilities of compounds A^3B^5 and isoelectronic elements of group IV may change with the position of the isoelectronic series in the periodic system.

In the upper part of the system, for example, in the silicon series, the increase in ionic character takes place more rapidly than in the lower part, for example in the isoelectronic series of grey tin. This can be ascertained from the difference between the electron affinities of the elements (such a rough approximation can be applied only to isoelectronic series). It is also indicated by the difference, which increases from the bottom to the top, in the forbidden gap widths of the corresponding (adjoining) compounds A^4 and A^3B^5. Thus, it is possible, in spite of the abrupt change in bonding from covalent to ionic, for the carrier mobility in aluminium phosphide to be no greater than that in silicon.

Prediction from Gubanov's theory, of the magnitude of the carrier mobility in gallium arsenide, was confirmed [200]. This quantity is only twice as great as the mobility in germanium, whereas indium antimonide has a mobility dozens of times greater than that of grey tin.

In connection with all the above, special importance is assumed by more complete isoelectronic series which also include ternary compounds. Isoelectronic series consisting of excess or defect phases are also interesting and have not, as yet, been studied.

3. Correlation of the Properties of Diamond-like Semiconductors

As was apparent from Parts 1 and 2, the physicochemical and electrical properties of diamond-like substances are related to the position of their constituent elements in the periodic system. It may be observed that in the group of substances under consideration there is a sharing of the chemical and physical properties that is based on the sharing of a basic type of chemical bond. The regular variation of the properties of substances with increasing atomic weight in series of analogues and the variation of the properties in isoelectronic series, are determined by the position of the constituent elements in the periodic system.

The scientific prediction of the properties of substances still unknown derives from these general assumptions. More detailed predictions necessitate the establishment of correlations between the various parameters of substances. These correlations are particularly interesting in cases where they relate an easily determinable or already known parameter of the substance to such basic electronic properties as the forbidden gap width (activation energy of intrinsic conduction) and the carrier mobility, which are of fundamental importance in selecting semiconductors for practical applications.

The group of diamond-like semiconductors, which is most interesting for radio electronics and has therefore been studied most extensively, has been the main object of such correlations.

The ideas of Welker, who correlated the forbidden gap width and the carrier mobility with the bond energy, were discussed above. However, they were too general to permit a correct prediction of the properties of semiconductors which have not yet been studied.

In the work of Zhuze[583], correlations are established for the degree of ionic character of a semiconductor as determined from the electronegativity, the forbidden gap widths and mobilities, and the functional dependence between the electronic mobility and the heat of formation within a group of semiconductors of a single structural type, particularly for semiconductors with the sphalerite structure. However, the method of evaluating the bond strength from that heat of formation is not applicable to substances with a pronounced covalent bond, as was pointed out by Ormont[584]. In the latter's view, in a group of substances A^4, A^3B^5 and A^2B^6, the forbidden gap width is related to the energy of atomic dissociation and to the specific surface energy, and he proposes the following equations[203]:

$$\Delta E = \left(\frac{v_B}{v_A}\right)^n \left[C - M + P\right] \Sigma_{hkl} \qquad (1)$$

$$\Delta E = \left(\frac{v_B}{v_A}\right)^n \left[c - m + p\right] \Omega. \qquad (2)$$

The term $\frac{v_B}{v_A}$ is the ratio of the effective charges of the atoms constituting the substances, the terms M and m are functions of the total atomic number of these atoms, the terms P and p are functions of the difference in electronegativities; Σ_{hkl} is the specific surface energy, Ω is the energy of atomic dissociation of compound AB into the ground state, and C and c are constants. It is arbitrarily assumed that the relative values of C, M and P and their signs reflect the influence of the covalent, metallic and ionic components of the chemical bond.

The values of the terms, factors and of the exponent n were found to be the same for the series A^4, A^4B^4, and A^3B^5. The ΔE values of the substances A^4, A^4B^4, A^3B^5 and A^2B^6 were thus calculated. The agreement between the calculated and the experimental values is good in a number of cases. However, the forbidden gap width of boron phosphide which has now been determined, differs from the predicted value of 1·4 eV. It appears that the forbidden gap widths of certain other substances (e.g., indium arsenide, gallium phosphide, etc.) do not obey the above-mentioned relations either.

Goodman[585] considered that the distances between the atoms of elements of subgroup IVb is an equilibrium distance characteristic of the "neutral" covalent bond in compounds A^3B^5 as well. He established a linear dependence between ΔE and $1/d^2$, where d is the inter-atomic distance. The extent of deviation from this dependence determines the ionic contribution into the forbidden gap width in binary compounds with the sphalerite structure.

Combining the interatomic distances found experimentally

(which are undoubtedly the result of a balancing of all types of interaction) with the forbidden gap widths, Goodman obtained an approximate agreement between the mobility and the degree of ionic character.

Folberth[586], analysing the relative values of the forbidden gap width and carrier mobility in compounds of type A^3B^5, reached the conclusion that in order to predict these values it is necessary to consider the deformation of the electronic structure, i.e., the polarization of the valence electrons in the direction of the atoms B^5. The polarization increases with an increase of the difference in the electronegativities of the atoms B^5 and A^3 and with an increase of the atomic weights. If, because of the polarization, each atom of B^5 is associated with an average of five electrons, and each atom of A^3 with three electrons, the difference between the "potential wells" tends to zero. A further rise in polarization causes the potentials of the conduction electrons to be greater relative to the atoms of A^3 than relative to the atoms of B^5, i.e., the sites occupied by the atoms of A^3 are more negatively charged than the sites occupied by the atoms of B^5. This overcharging of the sites, or the "anti-ionic state", which results from Folberth's bond model, does not correspond to the experimental data of Bäuerlein[580].

The latter showed that the energy of formation of Frenkel defects in compounds of type A^3B^5 is less than that in silicon and germanium, indicating the ionic character of the bond in these compounds. Moreover, it was found that the energy of formation of a defect at the site of arsenic is greater than that at the site of gallium (for gallium arsenide). The latter fact constitutes a direct refutation of Folberth's model.

In the latest work of the same author, for example in reference 587, the initial state taken is purely ionic, and the deformation of the anionic electron shells is taken into account, i.e., the author returns to earlier views of the covalent bond model.

However, neither Folberth's method (in both variants) nor Goodman's method, which was discussed before, can account for or predict the relative values of the forbidden gap width and carrier mobility in groups of diamond-like semiconductors.

An attempt at a more general approach of this type is described by Pearson[588]. The directionality of the bonds in the compound is estimated by means of the average value of the principal quantum numbers of the valence shells of the atoms composing the compound. The polarization is measured by the radius ratio of cation to anion, combined with the difference in electronegativity.

When ΔE is represented as a function of n and Δx in the space of the coordinates $n - \Delta x - r_c/r_a$, values of ΔE are obtained with an accuracy of 3 per cent, but only for certain compounds. Pearson finds his attempt to find a quantitative relation between the forbidden gap width and the three chosen parameters ($\bar{n}$, Δx, and r_c/r_a), unsuccessful, but holds that the latter are sufficient to provide a qualitative picture of the phenomenon.

A study has recently appeared on the relation between the bond energy and the forbidden gap width in diamond-like semiconductors [589]. The author found that a linear dependence exists between these quantities which for semiconductors of types A^4, A^3B^5 and A^2B^6 can be expressed as

$$\Delta E = a\,(\varepsilon_s - b),$$

where a and b are constants characteristic of each type of substance, and ε_s is the energy of a single bond.

ε_s is calculated in the manner of Pauling, taking into account the Hardy and Thomas electronegativities of the atoms; i.e., it represents a fairly rough estimate of the magnitude of the bond energy.

A correlation between hardness and interatomic distance was first noted by Goldschmidt[3]. He showed that $H = \text{const} \cdot r^{-m}$, where m is a quantity which is constant for groups of similar substances. In reference 49 it was established that $m = 9$ for compounds of type A^3B^5.

The problem of finding a correlation between microhardness and electrical properties was first stated in reference 204. Later, in reference 49, on the basis of an approximate correlation between the Knoop microhardness and the forbidden gap width, the conclusion was reached that the degree of ionic character decreases gradually in the series

AlP, GaP, AlAs, AlSb, GaAs, InP, GaSb, InAs, InSb.

Subsequently, microhardness as a quantity related to the energy of atomic dissociation, bond strength, bond character and other characteristics was used in many studies, but no quantitative conclusions were reached. Thus, a correlation was established in reference 203 between the microhardness and the specific surface energy of diamond-like substances.

As is evident from the examples cited, attempts to relate the energy, crystallochemical, mechanical and other properties theoretically to the electronic properties of diamond-like semiconductors have not yet produced satisfactory results, although a number of relationships makes it possible to evaluate the properties of still-unstudied substances to a first approximation.*

However, in our opinion, success in predicting the properties of semiconductors awaits the researcher not on the road leading to correlations between properties known at the present time, but along the path of further experimental investigations of the chemical bond (distribution of electron density, of effective charges, etc.) to be conducted together with complex exploratory studies.

---

* While this book was being proofread, Sucher's (sic) monograph *Physical Chemistry of Semiconductors* was published (Paris, 1962), in which a considerable amount of space is devoted to the relationship between the electronic and physicochemical properties of semiconductors.

CONCLUSION

ANALYSIS of the experimental material on diamond-like semiconductors shows that the bulk of the potentialities inherent in the properties of the substances in this group not only have not been exploited but are still largely unexplained. However, the analysis of the properties of known semiconductors and the observed regularities in the variation of their parameters permit the assumption that further multiple investigations of their chemical, physicochemical and electronic characteristics will reveal new combinations of these characteristics, so that it will be possible to develop new semiconductor devices and systems.

Photo-electronics, the principal field of application of semiconductors of the diamond group in a number of problems connected with the conversion of solar into electrical energy, requires that the mobility of the charge carriers be increased in the materials at the optimum forbidden band width (to obtain a large current for a given value of light absorption). An increase in the mobility of the charge carriers is also important in raising the high-frequency sensitivity of transistors, where the use of a material of higher carrier mobility gives a more rapid propagation of the injected impulse. These requirements interfere with the tendency towards raising the operating temperature of transistors, which is associated with a definite value of the width of the forbidden band of the material. As this width increases, the operating temperature rises, but the mobility drops. All these requirements are limited by the physicochemical properties of the materials, and cannot be changed by any improvements in the design of the devices.

As can be seen from the material presented above, subgroups of diamond-like semiconductors which are analogues of, for example, A^4, A^3B^5, etc., are examples of materials where the combinations of the basic physicochemical and electronic properties are confined

CONCLUSION

within definite limits. For instance, in subgroup A^4, including silicon–germanium solutions, the intrinsic width of the forbidden band, the mobility of the majority current carriers and the melting point are such that if we wish to obtain a width of the forbidden band of more than 1 eV, we shall necessarily obtain majority carrier mobilities of less than 2000 cm^2/V sec and shall deal with substances melting at temperatures above 1200°C. In compounds of type A^3B^5, this combination of the main parameters is much more efficient from the standpoint of the requirements mentioned above. For example, in gallium arsenide it is possible to obtain, for the same melting point of the material, a width of the forbidden band that is 1·5 times larger and a majority carrier mobility that is over 3 times as great.

In other groups, these combinations are different. For instance, in ternary compounds of type $A^2B^4C_2^5$, judging from the substance CdSnAs$_2$ (see Part D), the combination of width of the forbidden band, melting point and mobility differs markedly from the combination of these parameters in compounds of type A^3B^5.

Substitutional solid solutions of substances of the diamond-like group will apparently be the basic materials used in the construction of devices utilizing a variable width of the forbidden band.

Substances having a high carrier mobility and a high sensitivity to rays in the infrared portion of the spectrum will be used as materials for devices detecting the presence of heated bodies. This important technological field also requires materials of the diamond-like group, but the potentialities involved here have been far from exhausted (such substances consist of atoms in the seventh and sixth period of the periodic table).

The development of engineering involving the use of high temperatures introduces the problem of creating high-temperature rectifiers and sensing elements. Semiconductors with limiting temperatures of 400 to 600°C cannot be used for such purposes; much more refractory materials are required.

Among binary diamond-like compounds there are a great many refractory materials which are very promising in this respect and which surpass such a comparatively refractory material as silicon. Such compounds are silicon carbide, gallium phosphide, and also the phosphide and arsenide of boron; these substances are stable up to very high temperatures in various media.

An extensive opportunity for varying the physicochemical and electronic properties is afforded by solid solutions based on isovalent and, particularly, heterovalent substitution. The latter type of solution may find applications not only in the fields mentioned above, but also as materials for the direct conversion of thermal energy into electrical energy. For this, the materials should have a high thermoelectromotive force, a high electrical conductivity, and a low thermal conductivity.

As already stated above, solid solutions based on isovalent substitution show a tendency towards a considerable decrease in thermal conductivity. Apparently, this tendency will be substantially greater for heterovalent substitution. Those areas, and also others where semiconductors have already been or will be applied, such as television, the utilization of the energy of radioactive decay, the generation of radio waves, the control of automation, catalysis in chemical processes and many, many others, will require a rational selection of materials with the optimum combination of properties for each individual area of technical application. This problem cannot be solved in any way other than by the multiple investigation of large groups of substances related by similarities in chemical structure.

Summing up the results obtained from the experimental material on the group of diamond-like semiconductors, we come to the conclusion that the number of semiconductor materials, until recently limited, can today be regarded as truly infinite in this group alone. The properties and combinations of properties of the various semiconductors in this group are extremely varied.

This situation developed as a result of the exploration of tetrahedral phases more complex than binary ones, of which the earliest were solid solutions based on isovalent substitution. These tetrahedral phases of variable composition served to confirm an experimentally important idea underlying the investigation of complex semiconductor systems. The idea had as its original source the hypothesis put forth by Goldschmidt that in substances of tetrahedral structure the "individuality of the separate structural units recedes to the background before the entire structure of the crystal as a whole"[3].

Ideas developed later about the determining role of the electronic lattice in diamond-like substances provided the basis for the prediction that, despite the increase in the number of the components (kinds of atoms at the lattice points), the basic properties of a given group of semiconductors are preserved in multicomponent phases. This applied primarily to the mobility of the current carriers.

On the basis of the example of solid solutions between compounds of type A^3B^5 having maximum carrier mobilities, it was shown that in intermediate compositions the values of this parameter were close to the arithmetic mean of the values for the initial binary compounds. It was found that the scattering of electrons in the disordered structure of the alloy (as a result of the statistical substitution of one kind of atom for another) makes a negligible contribution to the overall mechanism of electron scattering. Moreover, three-component alloys of this type retain all the characteristics of the electronic properties of compounds of type A^3B^5. However, in these cases the atoms involved in the substitution were similar in nature and belonged to the same group of the periodic table (the systems InSb – GaSb, AlSb – GaSb, etc.).

These findings still left room for doubt as to the possibility of obtaining substances with properties similar to those of A^4 and A^3B^5 in more complex alloys (than binary compounds) whose composition included atoms of different groups of the periodic system. The discovery that the carrier mobility in $CdSnAs_2$ was just as high as in its electronic analogue, indium arsenide, dispersed these doubts once and for all.

The promise of multicomponent heterovalent phases as materials for radio electronics has now become obvious, and the paths traced for their exploration will probably soon become as well trodden as those followed in the exploration of binary tetrahedral phases.

The discovery of the high mobility in $CdSnAs_2$ showed, in addition, that high carrier mobility can be observed in substances with the structure of chalcopyrite, a fact which was formerly rejected by some investigators.

The appearance of data on the electronic properties of Cd_3As_2 necessitated a revision of the concept of the group of normal tetrahedral phases, which were believed earlier to be the only ones capable of having a high carrier mobility. It now appears that the excess tetrahedral phases, which have been studied but little as semiconductors, have properties that are no less and perhaps more interesting for semiconductor engineering than the properties of normal tetrahedral phases.

However, a comparison of cubic and non-cubic structures, as was done in one of the first publications on the electronic properties of Cd_3As_2, turned out to be irrelevant, while the treatment of defect, normal and excess tetrahedral phases from the standpoint of the gradual filling of vacancies in the closer packing* of the non-metal apparently does show promise. From this standpoint, Cd_3As_2 is a member of the crystallochemical group of excess tetrahedral phases.

Among the above-described phases there are also those whose electronic properties are intermediate between those of typical tetrahedral phases and those of semiconductor glasses: the defect

---

* The term "closer packing" is used arbitrarily in this case: as we know, the structures of diamond and its analogues are examples of "loose packing".

substances. Their investigation has successfully combined two paths: the exploratory path and one directed toward the construction of a detailed picture of the physiocochemical and electronic properties of individual substances. Their semiconducting properties were predicted and experimentally demonstrated soon after the first X-ray diffraction studies in the course of the exploratory investigations.

A detailed study of the properties of one of these substances showed the part played by controlled structural defects in the electronic properties, and also the part played by the ordering of the defects. Meanwhile, further exploratory studies led to the creation of a system of defect tetrahedral phases (see Table 10), where the number of defects in the structure can change between very wide limits, as can the number and nature of the constituent atoms. On the basis of such a system, it is possible to select a substance and to predict approximately the manner in which the electronic properties related to the defectiveness and ordering of the structure would be affected.

Recent work in the field of investigation of ternary systems containing tetrahedral phases has led to interesting results: it was found that the formation of tetrahedral phases was possible, where one and the same element may be assumed to have variable or different valences (solutions of elements of group IV in ternary compounds, solutions of InTe in InAs, etc., possessing the structure of sphalerite). These phases constitute a link between two different crystallochemical groups and are of great scientific interest.

The development of research in semiconductor compounds corresponds to the trend in modern chemistry which tries to progress from the periodic law, which ties together the system of elements, to more complex relationships underlying the systems of chemical compounds.

The expansion of research in the field of directed syntheses of semiconductors, the finding of relationships between groups of semiconductors, and the discovery of the laws governing their formation and relating the properties with the composition, constitute a concrete course to be followed in the development of semiconductor chemistry, and one which derives completely from the preceding development of ideas on the periodicity of properties and chemical structure.

REFERENCES

1. A. F. Ioffe, *Izv. AN SSSR, ser. fiz.*, **15**, 477, 1951.
2. R. L. Myuller, *Byulleten' Vsesoyuzn. khim. obshch. im. Mendeleeva*, **6**, 12, 1939.
3. V. M. Goldschmidt, *Usp. fiz. nauk*, **9**, 6, 811, 1929.
4. N. A. Goryunova. In: *Problems of the Theory and Investigation of Semiconductors and Processes of Semiconductor Metallurgy*.M., Izd. AN SSSR, 1955, p. 29.
5. N. A. Goryunova, *Izv. AN SSSR, ser fiz.*, **21**, 1, 120, 1957.
6. H. Hahn, G. Frank, W. Klingler, A. D. Störger, G. Störger. *Zs. anorg. allg. Chemie*, **279**, 5-6, 241, 1955.
7. B. F. Ormont, *Structure of Inorganic Substances*. M.-L., Gostekhizdat, 1950, p. 168.
8. J. Kimball. In: *Semiconducting Substances* (translation into Russian). M., IL., 1960, p. 11.
9. E. Mooser, W. B. Pearson. *Acta cryst.*, **12**, 1015, 1959.
10. W. B. Pearson. In: *Semiconducting Substances* (Translation into Russian). M., IL., 1960, p. 251.
11. B. V. Nekrasov, *Textbook of General Chemistry*. M., Goskhimizdat, 1953, p. 849.
12. L. H. Ahrens. *Geochim. cosmochim. acta*, **3**, 1, 1953.
13. D. Morris, L. Ahrens. In: *Semiconducting Substances*, (translation into Russian). M., IL, 1960, p. 57.
14. V. M. Goldschmidt. *Geochemistry*. Oxford. 1954.
15. H. G. Grimm, A. Sommerfeld. *Zs. Phys.*, **36**, 36, 1926.
16. C. H. L. Goodman. *J. Phys. Chem. Solids*, **6**, 4, 305, 1958.
17. I. D. Borneman-Starynkevich, *Dokl. Akad. Nauk SSSR*, **19**, 255, 1938.
18. N. A. Goryunova, C. S. Grigor'eva, *Z. tekh. fiz.*, **26**, 10, 2157, 1956.
19. N. A. Goryunova, *Vestnik Len. gos. univ.*, **10**, 112, 1961.
20. N. A. Goryunova, V. D. Prochukhan, *Fiz. tverd. tela*, **2**, 1, 176, 1960.
21. N. A. Goryunova, A. A. Vaipolin, Chiang Ping-hsi, In: *Physics and Chemistry*. L., Izd. LISI, 1961, p. 26.
22. N. A. Goryunova, A. V. Voitsekhovskii, V. D. Prochukhan, *Vestnik Len. gos. univ.*, **10**, 2, 156, 1961.

23. V. D. Prochukhan, Master's dissert. L., 1961.
24. E. E. Loebner, I. Y. Hegyi, E. W. Poor (Preprint).
25. E. Mooser, W. B. Pearson. *Phys. rev.*, **101**, 492, 1956.
26. Ya. A. Ugai. In: *Problems of the Metallurgy and Physics of Semiconductors*. M., Izd. AN SSSR, 1961, p. 107.
27. E. Mooser, W. B. Pearson. In: *Semiconducting Substances* (Translation into Russian). M., IL, 1960, p. 183.
28. N. A. Goryunova, A. A. Vaipolin, *Abstracts of the 4th Conference on Crystal Chemistry*. Kishinev, Izd. Mold. AN, 1961, p. 46.
29. R. Juza, W. Schulz. *Zs. anorg. Chem.*, **275**, 65, 1954.
30. R. Juza, F. Hund. *Zs. anorg. Chem.*, **257**, 13, 1948.
31. R. Juza, H. Weber, E. Meyer-Simon. *Zs. anorg. Chem.*, **273**, 48, 1953.
32. B. Renter, K. Hardel. *XVII Intern. Kongr. München. Kurzreferaten*, 155, 1959.
33. Y. A. A. Ketelaar. *Zs. Krist.*, **87**, 436, 1934.
34. G. Busch, E. Mooser, W. B. Pearson. *Helv. phys. acta*, **29**, 192, 1956.
35. H. Hahn. *XVII Intern. Kongr. München. Kurzreferaten*, 157, 1959.
36. A. Weiss, H. Schäfer. *Naturwissenschaften*, **47**, 955, 1960.
37. *Silicon. Collection of Translations* edited by D. A. Petrov. M., IL, 1960.
38. *Germanium. Collection of Translations* edited by D. A. Petrov. M., IL, 1955.
39. *Technology of Semiconducting Materials*. M., Oborongiz, 1961.
40. B. A. Krasyuk, A. I. Gribov, *Semiconductors. Germanium and Silicon*. M., GINTL, 1961.
41. V. P. Zhuze, *Semiconducting Materials*. L., Izd. Len. Doma nauch.-tekh. prop., 1957.
42. W. C. Dunlap, *An Introduction to Semiconductors* (Translation into Russian). M., IL, 1959.
43. A. F. Ioffe, *Radiotekhnika i elektronika*, **1**, 8, 1036, 1956.
44. A. F. Ioffe, *Physics of Semiconductors*. Izd. AN SSSR, 1957.
45. Coll. *Semiconductors in Science and Technology*. M., L., Izd. AN SSSR, 1957.
46. G. S. Zhdanov, *Solid State Physics*. Izd. Moskov. gos. univ., 1961, p. 119.
47. W. Kauzmann, *Quantum Chemistry: An Introduction* (Translation into Russian). IL, 1960, p. 312.
48. G. B. Bokii, *Introduction to Crystal Chemistry*. Izd. Moskov. gos. univ., 1954, pp. 102, 160, 218.
49. G. A. Wolff, L. Toman, N. J. Field, Y. C. Clare. *Halbleiter und Phosphore*. Braunschweig, 463, 1958.
50. G. A. Wolff, J. D. Broder. *Acta crystallographica*, **12**, 4, 313, 1959.
51. Selected values of chemical thermodynamic properties. *Circular of the National Bureau of Standards* 500. Washington, 1952.
52. T. S. Moss, *Photoconductivity in the elements*. London, 1952.
53. E. W. J. Mitchell. *J. phys. chem. solids*, **8**, 444, 1959.
54. R. H. Wentorf, H. P. Bovenkerk. *J. chem. phys.*, **36**, 8, 1987, 1962.

REFERENCES

55. A. Neuhaus. H. J. Meyer. *Angew. chem.*, **69**, 17, 551, 1957.
56. L. F. Grigoreva, Master's dissert. L., 1960.
57. A. Levitas, C. C. Wang, B. H. Alexander. *Phys. rev.*, **95**, 846, 1954.
58. H. D. Erfling. *Ann. Phys.*, **41**, 467, 1942.
59. M. Creen, J. A. Kafalus, P. H. Robinson. *Semiconducting surface physics*, 349, 1957.
60. A. J. Rosenberg. *J. Amer. chem. soc.*, **78**, 2429, 1956.
61. M. Hansen, K. Anderco. *Constitution of binary alloys.* Mc. Graw-Hill, 1958.
62. B. I. El'kin. In: *Problems of the Metallurgy and Physics of Semiconductors.* M., Izd. AN SSSR, 1957, p. 142.
63. A. R. Regel', Doctoral dissert. L., 1956.
64. D. A. Petrov, *Izv. AN SSSR, ser. tekh.*, **11**, 82, 1956.
65. G. W. Green, C. A. Hogarth, F. A. Johnson. *J. electronics and controls*, **3**, 171, 1957.
66. W. L. Bond, W. Kaiser. *J. phys. chem. sol.*, **16**, 44,
67. *Silicon carbide, the high temperature semiconductor*, Pergamon Press, 1960.
68. K. Zückler. *Halbleiterprobleme*, **3**, 207, 1956.
69. *Handbook of chemistry and physics*, 37-th ed., 1955–1956.
70. Silicon carbide. *Proceedings of Conference.* Boston, 1959. Published, 1960.
71. C. Fritzsche. *Herstellung von Halbleitern.* Berlin, VEB Verlag Technik, S. 50, 1960.
72. J. H. Racette. *Phys. rev.*, **107**, 1542, 1957.
73. M. S. Ablova, *Fiz. tverd. tela*, **3**, 6, 1815, 1961.
74. G. C. Kuczynski, R. F. Hochman. *Phys. rev.*, **108**, 4, 946, 1957.
75. S. I. Novikova, *Fiz. tverd. tela*, **2**, 1, 43, 1960.
76. R. E. Honig. *R. C. A. rev.*, **28**, 195, 1957.
77. E. M. Pell. *J. phys. chem. sol.*, **3**, 74, 1957.
77. E. M. Pell. *J. phys. chem. sol.*, **3**, 74, 1957.
78. R. L. Myuller, A. V. Danilov, T. P. Markova et al., *Vestnik Len. gos. univ.*, 1960, 4, 80.
79. R. L. Myuller, N. A. Baglai, *Vestnik Len. gos. univ.*, 1960, 4, 88.
80. E. A. Efimov, I. T. Erusalimchik, *Zh. fiz. khim.*, **32**, 413, 1103, 1967, 1958; **33**, 441, 1959.
81. Yu. A. Pleskov, *Khimicheskaya nauka i promyshlennost*, **3**, 4, 113, 1958.
82. M. C. Cretella, H. C. Gatos. *J. electr. soc.*, **105**, 487, 1958.
83. B. I. Boltaks, *Diffusion in Semiconductors*. M., Fizmatgiz, 1961.
84. B. P. Konstantinov, L. A. Badenko, *Fiz. tverd. tela*, **2**, 11, 2696, 1960.
85. S. G. Kalashnikov, A. K. Mednikov, *Fiz. tverd. tela*, **2**, 2058, 1960; **3**, 224, 1961.
86. H. Stöhr, W. Klemm. *Zs. anorg. allg. Chem.*, **241**, 305, 1939.
87. C. D. Thurmond. *J. chem. phys.*, **57**, 827, 1953.
88. I. N. Belokurova, M. G. Kekua, D. A. Petrov, A. D. Suchkova, *Izv. AN SSSR, otd. tekh. nauk, Metallurgiya i toplivo*, **1**, 9, 1959.

89. B. G. Zhurkin, M. G. Kekua, I. N. Belokurova, *Trudy Inst. metallurgii im. A. A. Baikova*, **5**, 178, 1960.
90. D. A. Petrov, *Zh. fiz. khim.*, **21**, 12, 1449, 1947.
91. R. Klement, H. Sandmann. *Naturwissenschaft*, **44**, 349, 1957.
92. R. A. Swalin. *Acta metallurgia*, **6**, 2, 126, 1958.
93. *Atomes*, **10**, 110, 171, 1955.
94. F. Herman. *Phys. rev.*, **95**, 847, 1954.
95. B. Goldstein. *R. C. A. rev.*, **18**, 4, 458,-1957.
96. N. A. Goryunova, Master's dissert. L., 1950.
97. O. N. Tufte, A. W. Ewald. *Bull. Amer. phys. soc.*, **2**, 3, 128, 1958.
98. S. I. Novikova, *Fiz. tverd. tela*, **2**, 9, 2341, 1960.
99. N. N. Sirota, *Dokl. Akad. Nauk SSSR*, **51**, 4, 1946.
100. A. W. Ewald. *J. appl. phys.*, **25**, 11, 436, 1954.
101. M. M. Chertok, *Zh. tekh. fiz.*, **5**, 711, 1935.
102. A. I. Blum, N. A. Goryunova, *Dokl. Akad. Nauk SSSR*, **75**, 367, 1950.
103. G. Bush, I. Wieland, H. Zoller. In: *Semiconductor Materials* (Translation into Russian). IL, 1954.
104. J. T. Kendall. *Phil. mag.*, **45**, 141, 1954.
105. P. R. Kamadziew. *Ceskoslov. casop. fys.*, **4**, 424, 1954.
106. A. W. Ewald. *Phys. rev.*, **91**, 244, 1953.
107. E. E. Kohnke, A. W. Ewald. *ibid.* **97**, 607, 1955; **100**, 1251, 1955.
108. N. A. Goryunova, *Dokl. Akad. Nauk SSSR*, **75**, 1, 51, 1950.
109. N. A. Goryunova, N. N. Fedorova, V. I. Sokolova, *Zh. tekh. fiz.*, **28**, 8, 1672, 1958.
110. G. V. Ozolin'sh, G. K. Averkieva, A. F. Ievin'sh, N. A. Goryunova, *Abstracts of the 4th Conference on Crystal Chemistry*. Kishinev, Izd. Mold. AN, 1961, p. 108.
111. K. Smirous, *Czechosl. Journal of Physics*, **5**, 4, 537, 1955.
112. T. C. Herman. *J. electrochem. soc.*, **103**, 128, 1956.
113. D. N. Nasledov, A. Yu. Khalilov, *Zh. tekh. fiz.*, **36**, 6, 1956.
114. L. Pincherle, J. M. Radcliffe. *Adv. phys.*, **5**, 19, 271, 1956.
115. H. Welker. *Intern. conf. on semiconductor physics*. Prague, 889, 1960. publ. 1961.
116. R. Bowers, K. W. Ure, J. E. Bauerle, A. J. Cornish. *J. appl. phys.*, **30**, 6, 930, 1959.
117. A. V. Ioffe, A. F. Ioffe, *Fiz. tverd. tela*, **2**, 781, 1960.
118. N. A. Goryunova, *Zh. Vsesoyuz. khim. obshch. im. D. I. Mendeleeva*, **5**, 5, 522, 1960.
119. N. A. Goryunova, L. V. Kradinova, V. I. Sokolova, E. V. Sokolova, *Zh. fiz. khim.*, 33, 1409, 1960.
120. A. S. Pashinkin, A. A. Men'kov, A. V. Novoselova, *Zh. nauch. khim.*, **2**, 4, 826, 1957.
121. V. I. Likhtman, B. M. Maslennikov, *Dokl. Akad. Nauk SSSR*, **17**, 1, 93, 1949.
122. W. G. Pfann. *J. met.*, **4**, 747, 861, 1952.
123. B. I. Aleksandrov, B. E. Verkin, B. G. Lazarev, *Fizika metallov i*

REFERENCES

metallovedenie, **2**, 93, 1956.
124. O. G. Folberth. *Zs. Metallkunde*, **49**, 11, 570, 1958.
125. A. Addamiano. *J. Amer. chem. soc.*, **82**, 7, 1937, 1960.
126. G. Wolff, P. H. Keck, J. D. Bröder. *Bull. Amer. phys. soc.*, **29**, 1, 16, 1954.
127. R. Pease. *Acta cryst.*, **5**, 356, 1952.
128. R. H. Wentorf. *J. chem. phys.*, **26**, 956, 1957.
129. M. G. Valyashko, *Boron, Its Compounds and Alloys*. Kiev, Izd. AN USSR, 1960, p. 209.
130. E. Podszus. *Zs. anorg. Chem.*, **30**, 156, 1917.
131. G. A. Meerson, G. V. Samsonov, N. Ya. Tseitlina, *Ogneupory*, **20**, 72, 1955.
132. N. F. Zhirov, *Luminophors*. M., Gizoboronprom, 1940, p. 351.
133. K. Taylor. *Industr. engng. chem.*, **47**, 2506, 1955.
134. E. Brank, J. Margrave, V. Meloche. *J. inorg. nuclear chemistry*, **5**, 48, 1957.
135. A. V. Moskvin, *Cathodoluminescence*. M., L., Gostekhizdat, **2**, p. 515, 1948.
136. E. Wiberg, H. Mischand. *Zs. Naturf.*, **96**, 497, 1954.
137. S. Geller. *Phys. chem. solids*, **10**, 340, 1959.
138. R. Wickery. *Nature*, **184**, 268, 1959.
139. S. Rundgvist. XVI Congrés International de chimie pure et appliquée. Paris. *Memoire présenté a la section de chimie minerale*, p. 539, 1958.
140. P. Popper, T. A. Ingles. *Nature*, **179**, 1075, 1957.
141. B. Stone, D. Hill. *Phys. rev. letters*, **4**, 6, 282, 1960.
142. M. A. Besson. *Compt. rend. Paris*, **113**, 78, 1891.
143. M. H. Moissan. *Compt. rend. Paris*, **113**, 624, 726, 1891.
144. R. V. Williams, R. A. Ruherwein. *J. Amer. chem. soc.*, **82**, 6, 1330, 1960.
145. V. I. Matkovich. *Acta cryst.*, **14**, 93, 1961.
146. J. A. Perri, Sam La-Placa, B. Post. *Acta cryst.*, **11**, 310, 1958.
147. B. F. Ormont, *Zh. nauch. khim.*, **4**, 2176, 9, 1959.
148. M. V. Stackelberg, K. Spiess. *Zs. phys. Chem.*, A **175**, 127, 1935.
149. T. Renner. *Zs. anorg. Chemie*, **298**, 22, 1959.
150. R. Kauer, A. Rabenau. *Zs. Naturf.*, **12a**, 11, 942, 1957.
151. V. M. Goldschmidt. *Skrifter Norske Videnskop-Akademie Oslo, Math. nat.* 8, 1927.
152. M. P. Slavinskii, *Physicochemical Properties of the Elements*. M., GNTIL chern. tsvetn. metallurgii, 1952.
153. F. Herman. *J. electronics*, **1**, 103, 1955.
154. Yu. I. Pashintsev, Author's abstract of dissertation. Minsk, 1959.
155. H. G. Grimmes, W. Kischio, A. Rabenau. *J. phys. chem. sol.*, **16**, 302, 1960.
156. W. Köster, B. Thoma. *Zs. Metallkunde*, **46**, 291, 1955.
157. O. G. Folberth, F. Oswald. *Zs. Naturf.*, **9a**, 1050, 1954.
158. A. S. Borshchevskii, D. N. Tret'yakov, *Fiz. tverd. tela*, **1**, 9, 1483,

1959.
159. G. Giesecke, H. Pfister. *Acta cryst.*, **11** (5), 369, 1958.
160. A. V. Gul'tyaev, A. V. Petrov, *Fiz. tverd. tela*, **1**, 3, 368, 1959.
161. G. G. Urazov, *Izv. inst. fiz.-khim. Analiza*, **1**, 461, 1921.
162. W. Guertler, A. Bergmann. *Zs. Metallkunde*, **25**, 4, 81, 1933.
163. F. Sauerwald. *Zs. Metallkunde*, **14**, 468, 1920.
164. O. G. Folberth. *Halbleiterprobleme*, 5. Braunschweig, 1960.
165. V. M. Glazov, Diss., M., 1959.
166. R. Ward, R. Ray, A. Ulich. *Calculation reactions of metals*. California, 1948.
167. F. Oswald, K. Schade. *Zs. Naturf.*, **9a**, 611, 1954.
168. A. R. Regel', M. S. Sominskii, *Zh. tekh. fiz.*, **25**, 768, 4, 1955.
169. D. A. Petrov, M. S. Mirgalovskaya, I. A. Strel'nikova, E. M. Komova, *Problems of the Metallurgy and Physics of Semiconductors*. M., Izd. AN SSSR, 1957, p. 70.
170. H. A. Shell. *Zs. Metallkunde*, **49**, 140, 1958.
171. E. Justi, J. Lautz. *Abh. Braunschw. wissenschaft. Gesellschaft*, **5**, 36, 1953.
172. H. Welker, H. Weiss. In: *New Semiconducting Materials* (Translation into Russian). IL, 1958.
173. W. P. Allred, W. L. Meffered, R. K. Willardson. *J. electr. soc.*, **107**, 2, 117, 1960.
174. V. A. Presnov, V. F. Synorov, *Zh. tekh. fiz.*, **27**, 1, 123, 1927.
175. R. Juza, H. Hahn. *Zs. anorg. allg. Chem.*, **239**, 282, 1938.
176. G. S. Zhdanov, G. V. Mirman, *Zh. eksp. teor. fiz.*, **6**, 10, 1201, 1936.
177. H. Hahn, R. Juza. *Zs. anorg. allg. Chem.*, B **244**, 111, 1940.
178. D. Effer, G. R. Antell. *J. electr. soc.*, **107**, 110, 1960.
179. G. V. Samsonov, L. L. Vereikina, Yu. V. Titkov, *Zh. nauch. khim.*, **6**, 3, 749, 1961.
180. C. J. Frosch, L. Derick. *J. electrochem. soc.*, **103**, 2, 251, 1961.
181. G. Wolff, R. Herbert, J. D. Bröder. In: *New Semiconducting Materials* (Translation into Russian). M., IL., 1958, p. 201.
182. C. Hilsum, A. C. Rose-Innes. *Semiconducting III–V compounds*. N. Y. Pergamon Press, **1**, 1961.
183. N. N. Sirota, L. I. Berger, *I. fiz. zh.*, **1**, 11, 117, 1968.
184. S. I. Novikova, *Fiz. tverd. tela*, **3**, 1, 178, 1961.
185. J. van den Boomgaard, K. Schol. *Philips research reports*, **12**, 2, 127, 1957.
186. Landolt-Börnstein. *Phys. chem. Tabellen*. Berlin, Julius Spinger. 1912–1935.
187. A. S. Borshchevskii. In: *Physics and Chemistry*. L., Izd. LISI, 1961, 19.
188. G. K. Averkieva, O. V. Emel'yanenko, *Zh. tekh. fiz.*, **28**, 1945, 9, 1958.
189. D. N. Nasledov, A. Ya Patrakova, B. V. Tsarenkov, *Zh. tekh. fiz.*, **28**, 4, 779, 1958.
190. D. A. Jenny. *Proc. I. R. E.*, **46**, 6, 959, 1958.

REFERENCES

191. L. R. Weisberg, J. R. Woolston, M. Glickmans. *J. appl. phys.*, **29**, 10, 1514, 1958.
192. O. V. Emel'yanenko, D. N. Nasledov, *Zh. tekh. fiz.*, **28**, 6, 1177, 1958.
193. D. N. Nasledov, N. N. Smirnova, B. V. Tsarenkov, *Fiz. tverd. tela*, **2**, 2782, 11, 1960.
194. D. N. Nasledov, B. V. Tsarenkov, *Fiz. tverd. tela*, **1**, 1467, 1959.
195. D. N. Nasledov, B. V. Tsarenkov, *Fiz. tverd. tela*, **1**, 78, 1959.
196. R. Gremmelmaier, H. J. Henkel. *Zs. Naturf.*, **14**a, 1072, 1959.
197. K. Begner, K. Shmirous, *Czechosl. Journal of Physics*, **5**, 546, 1955.
198. W. F. Schottky, M. B. Bever. *Acta metallurgica*, **6**, 5, 320, 1958.
199. V. M. Glazov, A. A. Vertmen, *Dokl. Akad. Nauk SSSR*, **123**, 3, 492, 1958.
200. N. A. Goryunova, doct. dissert. L., 1958.
201. D. P. Detwiler. *Phys. rev.*, 97, 1575, 6, 1955.
202. R. Juza, A. Rabenau, G. Pascher. *Zs. anorg. allg. Chem.*, **285**, 242, 1956.
203. B. F. Ormont. In: *Problems of the Metallurgy and Physics of Semiconductors.* M., Izd. AN SSSR, 1961, p. 5.
204. A. S. Borshchevskii, N. A. Goryunova, N. K. Takhtareva, *Zh. tekh. fiz.*, **27**, 1408, 7, 1957.
205. A. Tiel, W. Koelsch. *Zs. anorg. Chem.*, **66**, 288, 1910.
206. M. Gliksman, K. Weiser. *J. electrochem. soc.*, **105**, 12, 728, 1958.
207. T. S. Liu, E. Peretti. *Trans. Amer. soc. metals*, **45**, 677, 1953.
208. E. Schillmann. *Zs. Naturf.*, **11**a, 6, 1956.
209. D. Effer. *J. electrochem. soc.*, **108**, 4, 357, 1961.
210. H. B. Gutbier. *Zs. Naturf.*, **14**[a], 210, 32–34, 1, 1959.
211. K. G. Günther. *Zs. Naturf.*, **13**[a], 211, 1081, 10, 1958.
212. H. Welker. *Scientia electrica*, **1**, 152, 1954.
213. A. G. Samoilovich, L. L. Kremblit, *Uspekh. fiz. nauk*, **57**, 4, 577, 1955.
214. V. P. Zhuze, A. G. Tsidil'kovskii, *Zh. tekh. fiz.*, **28**, 23, 2, 1958.
215. A. R. Regel', *Semiconductor Devices For Measuring Magnetic Field Strength.* Ser. *Semiconductors and Their Technical Applications.* L., Izd. Doma nauch-tekh. prop., 1956, p. 4.
216. V. P. Zhuze, A. R. Regel', *Technical Applications of the Hall Effect.* L., Izd. Doma nauch.-tekh. prop., 1957.
217. N. P. Mokrovskii, A. R. Regel', *Zh. tekh. fiz.*, **22**, 1281, 8, 1952.
218. S. N. Gadzhiev, K. A. Sharifov, *Dokl. Akad. Nauk SSSR*, **136**, 6, 1339, 1961.
219. S. A. Pogodin, S. A. Dubinskii, *Izv. sektora fiz.-khim. analiza*, **17**, 204, 1949.
220. A. A. Nesmeyanov, B. Z. Iofa, A. S. Polyakov, *Zh. khim.*, **5**, 2, 246, 1960.
221. J. D. Venables, R. M. Brondy. *J. appl. phys.*, **29**, 7, 1025, 1958.
222. R. E. Maringer. *J. appl. phys.*, **29**, 1261, 8, 1958.
223. M. C. Lavine, A. J. Rosenberg, H. C. Gatos. *J. appl. phys.*, **29**, 1131, 7, 1958.

224. M. C. Lavine, H. C. Gatos. *J. electrochem. soc.*, **107**, 5, 427, 1960.
225. A. J. Rosenberg. *J. phys. chem. sol.*, **14**, 175, 1960.
226. R. L. Myuller, G. M. Orlova, Ts'ui Ch'in Hua, *Zh. obshch. khim.*, **31**, 8, 2457, 1961.
227. R. L. Myuller, G. M. Orlova, Ts'ui Ch'in Hau, *Zh. obshch. khim.*, **31**, 8, 2461, 1961.
228. R. L. Myuller, T. P. Markova, S. M. Repinskii, *Vestnik Len. gos. univ.*, **16**, 3, 106, 1959.
229. S. A. Semiletov, *Trudy inst. Kristallografii*, **11**, 121, 1955.
230. S. A. Semiletov, M. Rozsival, *Kristallografiya*, **2**, 2, 287, 1957.
231. G. A. Kurov, Z. G. Pinsker, *Zh. tekh. fiz.*, **28**, 1, 29, 1958.
232. L. S. Palatnik, V. M. Kosevich, L. V. Tyrina, *Fizika metallov i metallovedenie*, **11**, 2, 229, 1961.
233. L. Reiner. *Zs. Naturf.*, **13ª**, 148, 1958.
234. J. B. Mullin. *J. electronics control*, **4**, 4, 358, 1958.
235. A. J. Strauss. *Bull. Amer. phys. soc.*, **3**, 2, 119, 1958.
236. K. F. Hulme, J. B. Mullin. *Phys. mag.*, **4**, 47, 1286, 1959.
237. W. G. Pfann. *J. metals*, **5**, 11, 1441, 1953.
238. K. I. Vinogradova, V. V. Galavanov, D. N. Nasledov, L. I. Solov'eva, *Fiz. tverd. tela*, **1**, 3, 403, 1959.
239. D. N. Nasledov, A. Yu. Khalilov, *Zh. tekh. fiz.*, **26**, 2, 251, 1956.
240. E. Burstein. *Phys. rev.*, **93**, 632, 3, 1954.
241. M. Haase. *Zs. Kristallograph.*, **65**, 509, 1927.
242. *Chemistry Handbook*. M. L., Goskhimizdat, vol. 1, 1951.
243. L. Moser, R.Erth. *Zs. anorg. allg. Chem.*, **118**, 269, 1921.
244. Ya. I. Gerasimov, A. N. Krestovnikov, *Chemical Thermodynamics in Nonferrous Metallurgy.* ONTI, No. 1, 1933.
245. S. Hartison. *Phys. rev.*, **93**, 52, 1, 1954.
246. B. Ya. Brach, V. V. Zhdanova, E. Ya. Lev, *Fiz. tverd. tela*, **3**, 3, 318, 1961.
247. A. S. Pashinkin, T. N. Tishchenko, I. V. Korneeva, B. N. Ryzhenko, *Kristallografiya*, **5**, 2, 261, 1960.
248. E. Krucheanu, Master's dissert. M., 1960.
249. S. A. Semiletov, *Kristallografiya*, **1**, 306, 3, 1956.
250. S. I. Novikova, *Fiz. tverd. tela*, **2**, 9, 2341, 1960.
251. I. V. Korneeva, Master's dissert., M., 1961.
252. F. A. Kröger, D. de Nobel. *J. Electronics*, **1**, 190, 1955.
253. F. A. Kröger, H. J. Vink, J. von den Boomgard. In: *Problems of Modern Physics*, issue *Luminescence* (Translation into Russian). IL, 1957.
255. R. Frerichs. *Phys. rev.*, **72**, 12, 1947.
256. W. W. Piper. *Phys. rev.*, **92**, 23, 1, 1953.
257. I. B. Mizetskaya, A. P. Trofimenko, V. D. Fursenko, *Zh. nauch. khim.*, **3**, 10, 2236, 1958.
258. M. E. Grillot. *Compt. rend.*, **242**, 779, 6, 1956.
259. E. T. Allen, J. L. Crenchow. *J. Amer. sci.*, **34**, 341, 1912.
260. A. Kremheller, A. K. Levine. *J. appl. phys.*, **28**, 746, 6, 1957.

REFERENCES

261. H. Fumeron-Rodot, M. Rodot. *Compt. rend.*, **248**, 937, 1959.
262. G. B. Dubrovskii, *Fiz. tverd. tela*, 3, 5, 1305, 1961.
263. R. J. Maurer. *J. chem. phys.*, **13**, 321, 8, 1945.
264. G. V. Hevesy. *Zs. phys. chem.*, **101**, 337, 1922.
265. K. Nagel, C. Wagner. *Zs. phys. Chem.*, **25**, 71, 1934.
266. C. Tuband. *Zs. anorg. Chem.*, **115**, 105, 1920.
267. J. W. Mellor. *A comprehensive treatise on inorganic and theoretical chemistry*, N. Y., 1946.
268. H. Welker. *Ergebnisse der exakten Naturwissenschaften*, **25**, 275, 1956.
269. C. Shih, E. A. Peretti, *J. Amer. chem. soc.*, **75**, 3, 608, 1953.
270. N. A. Goryunova, N. N. Fedorova, *Zh. tekh. fiz.*, **25**, 7, 1339, 1955.
271. O. G. Folberth. *Zs. Naturf.*, **10**, 6, 502, 1955.
272. N. A. Goryunova, V. I. Sokolova, *Izv. Moldav. filiala AN SSSR*, **3** (69), 97, 1960.
273. A. S. Borshchevskii, Master's dissert. L., 1961.
274. J. C. Woolley, B. A. Smith. *Proc. phys. soc.*, **70**, 1, 153, 1957.
275. N. A. Goryunova, I. I. Burdiyan, *Dokl. Akad. Nauk SSSR*, **120**, 5, 1031, 1958.
276. N. A. Goryunova, B. V. Baranov, *Dokl. Akad. Nauk SSSR*, **129**, 839, 4, 1959.
277. J. C. Woolley, B. A. Smith, D. G. Lees. *Proc. phys. soc.*, **69**, 12, 1339, 1956.
279. N. A. Goryunova, V. I. Sokolova, *Izv. Moldav. filiala AN SSSR*, **3** (69), 31, 1960.
280. O. G. Folberth, H. Welker. *J. phys. chem. solids*, **8**, 19, 14, 1959.
281. M. S. Abrahams, R. Braunstein, F. L. Rosi. *J. phys. chem. solids*, **10**, 204, 1959.
282. J. C. Woolley, C. M. Gillet, J. A. Evans. *Proc. phys. soc.*, **77**, 3, 700, 1961.
283. I. E. Gorshkov, N. A. Goryunova, *Zh. nauch. khim.*, **3**, 3, 668, 1958.
284. W. Köster, B. Thoma. *Zs. Metallkunde*, **46**, 4, 293, 1955.
285. I. I. Burdiyan, A. S. Borshchevskii, *Zh. tekh. fiz.*, **28**, 12, 2684, 1958.
286. I. I. Burdiyan, Master's diss. L., 1958.
287. A. S. Borshchevskii, I. I. Burdiyan, E. Yu. Lubenskaya, E. V. Sokolova, *Zh. nauch. khim.*, **4**, 12, 2824, 1959.
288. J. F. Miller, H. L. Goering, R. C. Himes. *J. electrochem. soc.*, **107**, 6, 527, 1960.
289. B. V. Baranov, N. A. Goryunova, *Fiz. tverd. tela*, **2**, 2, 284, 1960.
290. B. V. Baranov, Master's diss. L., 1961.
291. N. A. Goryunova, B. V. Baranov, V. D. Prochukhan, *Byulleten' izobretenii*, **6**, 1960, No. 127030.
292. V. M. Glazov, Liu Cheng-yüan, *Izv. AN SSSR, otd. tekh. nauk. Metallurgiya i toplivo*, **4**, 150, 1960.
293. A. A. Popov. In: *Problems of Metallurgy and Heat Treatment*. Sverdlovsk, 1956.
294. D. S. Kamenetskaya, *Zh. fiz. khim.*, **22**, 1, 81, 1948.
295. V. N. Romanenko, V. I. Ivanov-Omskii, *Dokl. Akad. Nauk SSSR*,

129, 3, 533, 1959.
296. Ya. Agaev, D. N. Nasledov, *Fiz. tverd. tela*, **2**, 5, 826, 1960.
297. Ya. Agaev, O. V. Emel'yanenko, D. N. Nasledov, *Fiz. tverd. tela*, **3**, 1, 194, 1961.
298. W. Köster, W. Ulrich. *Zs. Metallkunde*, **49**, 7, 1958.
299. J. S. Blakemore. *Canad. j. phys.*, **36**, 91, 1, 1957.
300. C. Kolm, S. Kulin, B. L. Averbach. *Phys. rev.*, **108**, 965, 1957.
301. P. Levesgue. *J. Metals*, **6**, 772, 1954.
302. B. P. Mitrenin, N. E. Troshin, K. P. Tsomaya, V. A. Vlasenko, Yu. D. Gubanov. In: *Problems of the Metallurgy and Physics of Semiconductors*. Izd. AN SSSR, 1957, p. 59.
303. V. I. Ivanov-Omskii, Master's dissert. L., 1961.
304. J. C. Woolley, C. M. Gillet. *J. phys. chem. solids*, **17**, 34, 1960.
305. N. A. Goryunova. *Proc. Intern. conference on semiconductor physics*. Prague, p. 909, 1960. Published in 1961.
306. N. A. Goryunova. In: *Problems of the Metallurgy and Physics of Semiconductors*. M., Izd. AN SSSR, 1961, p. 123.
307. V. S. Grigor'eva, Master's dissert. L., 1959.
308. H. Weiss. *Zs. Naturf.*, **11a**, 6, 430, 1956.
309. F. Oswald. *Zs. Naturf.*, **14a**, 4, 374, 1959.
310. O. V. Bogorodskii, A. Ya. Nashel'skii, V. Z. Ostrovskaya, *Kristallografiya*, **6**, 1, 119, 1961.
311. Nashel'slii, *Byulleten' izobretenii*, **12**, 1960, No. 129338.
312. *Electr. Manufakturer*, April 1958.
313. N. A. Goryunova, N. K. Takhtareva, *Izv. AN Moldav. SSR*, **10**, (88), 89, 1961.
314. G. B. Bokii, E. A. Pobedinskaya, *Vestnik Moskov. gos. uni.*, **3**, 121, 1955.
315. N. I. Vitrikhovskii, I. B. Mizetskaya, *Fiz. tverd. tela*, **2**, 10, 2579, 1960.
316. A. Kremheller, A. K. Livine, G. Gashurov. *J. electrochem. soc.*, **107**, 1 (1960).
317. N. A. Goryunova, V. A. Kotovich, V. A. Frank-Kamenetskii, *Zh. tekh. fiz.*, **25**, 2491, 14, 1955.
318. S. Forque, R. Goodrich, A. Cope. *R. C. A. rev.*, **12**, 335, 1951.
319. N. A. Goryunova, N. N. Fedorova, *Dokl. Akad. Nauk SSSR*, **90**, 6, 1039, 1953.
320. J. C. Woolley, B. Ray. *J. phys. chem. solids*, **13**, 151, 1960.
321. W. D. Lawon, S. Nielsen, E. H. Putley, A. S. Young. *J. phys. chem. solids*, **9**, 5, 1959.
322. A. D. Shneider, I. V. Gavrikhin, *Fiz. tverd. tela*, **2**, 9, 2079, 1960.
323. S. Larach, R. E. Schrader, C. F. Stocker. *Phys. rev.*, **108**, 3, 587, 3, 1957.
324. S. Larach, W. H. Carroll, R. E. Schrader. *J. phys. chem.*, **60**, 604, 5, 1956.
325. R. Yu. Khansevarov, S. M. Ryvkin, I. N. Ageeva, *Zh. tekh. fiz.*, **28**, 3, 480, 1958.

REFERENCES

326. N. I. Vitrikhovskii, I. B. Mizetskaya, *Fiz. tverd. tela*, **3**, 5, 1581, 1961.
327. E. N. Nikol'skaya, A. R. Regel', *Zh. tekh. fiz.*, **25**, 8, 1347, 1955.
328. K. Möonkemeyer. *Neues Jahre Min.*, **22**, 28, 1906.
329. T. Heumann. *Zs. Naturf.*, **5a**, 4, 216, 1950.
330. H. Hahn, G. Frank, W. Klingler, A. Meyer, Störger. *Zs. anorg. allg. Chem.*, **271**, 153, 1953.
331. V. P. Zhuze, V. M. Sergeeva, E. L. Shturm, *Zh. tekh. fiz.*, **28**, 10, 2093, 1958.
332. J. G. Austin, C. H. L. Goodman, A. E. Pengelli. *J. electrochem. soc.*, **103**, 609, 11, 1956.
333. V. M. Glazov, V. N. Vigdorovich, *Microhardness of Metals*, p. 199. M., Metallurgizdat, 1962.
334. J. H. Wernick, K. E. Benson. *J. phys. chem. sol.*, **3**, 157, 1957.
335. O. G. Folberth, H. Pfister. Halbleiter und Phosphore. Braunschweig, 474, 1958.
336. H. Pfister. *Acta crystallographica*, **11**, 221, 1958.
337. C. H. L. Goodman. *Nature*, **179**, 828, 1957.
338. O. G. Folberth, H. Pfister. *Acta crystallographica*, **13**, 199, 1961.
339. A. Rabenau, P. Eckerlin. *Naturwissenschaften*, **46**, 3, 106, 1959.
340. N. A. Goryunova, Chiang Ping-hsi, *Abstracts. All-union Conference on Semiconducting Compounds*. M., L., Izd. AN SSSR, 1961, p. 21.
341. A. J. Rosenberg, A. J. Strauss. *J. phys. chem. sol.*, **17**, 278, 1961.
342. N. A. Goryunova, S. Mamaev, V. D. Prochukhan, *Dokl. Akad. Nauk SSSR*, **142**, 3, 623, 1961.
343. V. I. Sokolova, *Abstracts of the All-Union Conference on Semiconducting Compounds*. M., L., Izd. AN SSSR, 1961.
344. N. A. Goryunova, G. K. Averkieva, Yu. V. Alekseev, *Vestnik Moldavskogo filiala AN SSSR*, **3**, 99, 1960.
345. J. H. Wernick. *US Patent*, **2**, 882, 192, Iss. 14, 1959.
346. A. T. Alieva, Z. G. Pinsker, *Kristallografiya*, **6**, 2, 204, 1961.
347. C. H. L. Goodman, B. W. Douglas. *Physica*, **20**, 1107, 1954.
348. N. A. Goryunova, N. N. Fedorova, *Fiz. tverd. tela*, **1**, 2, 344, 1959.
349. A. Addamiano. *J. electrochem. soc.*, **107**, 12, 1006, 1960.
350. A. V. Voitsekhovskii, N. A. Goryunova. In: *Physics*. LISI, 1962, p. 12.
351. N. A. Goryunova, V. S. Grigor'eva, P. V. Sharavskii, L. I. Osnach. In: *Physics*, L., Izd. LISI, 7, 1962.
352. N. A. Goryunova, G. K. Averkieva, P. V. Sharavskii, Yu. K. Tovpentsev. In: *Physics and Chemistry*. L., Izd. LISI, 22, 1961.
353. O. G. Folberth. In: *Semiconducting Substances* (Translation into Russian). M., IL, 1960, p. 213.
354. A. V. Voitsekhovskii, Abstracts of the All-Union Conference on Semiconductors. M, L., Izd. AN SSSR, 1961, p. 13.
355. S. Mamaev, *Izv. AN Turkm. SSR*, **6**, 7, 1960.
356. S. Mamaev, N. A. Nran'yan, *Izv. AN Turkm. SSR*, **5**, 21, 1961.
357. S. Mamaev, D. N. Nasledov, V. V. Galavanov, *Fiz. tverd. tela*, **3**, 11, 3405, 1961.
358. E. F. Apple. *J. electrochem. soc.*, **105**, 251, 1958.

359. H. Rodot. *Proc. Intern. Conference on semiconductor physics.* Prague 1960, publ. in 1961.
360. V. P. Chernyavskii, In: *Physics.* L., Izd. LISI, 1962, p. 10.
361. N. A. Goryunova, G. M. Orlova, A. V. Danilov, A. V. Abramova, R. L. Plechko, I. I. Kozhina, *Vestnik Len. gos. univ.*, **22**, 4, 97, 1961.
362. H. Hahn, W. Klingler. *Zs. anorg. Chem.*, **259**, 1-4, 121, 1949; **260**, 3, 97, 1949.
363. N. A. Goryunova, V. S. Grigor'eva, B. M. Konovalenko, S. M. Ryvkin, *Zh. tekh. fiz.*, **25**, 10, 1675, 1955.
364. S. A. Semiletov, *Fiz. tverd. tela*, **3**, 3, 746, 1961.
365. A. I. Zaslavskii, V. M. Sergeeva, I. A. Smirnov, *Fiz. tverd. tela*, **2**, 11, 2884, 1960.
366. E. Kauer, A. Rabenau. *Zs. Naturf.*, **13a**, 7, 531, 1958.
367. V. P. Zhuze, V. M. Sergeeva, A. I. Shelekh, *Fiz. tverd. tela*, **2**, 11, 2858, 1960.
368. J. Flahout. *Compt. rend.*, **232**, 334, 1951.
369. V. P. Mushinskii, *Fiz. tverd. tela*, **1**, 3, 515, 1959.
370. J. C. Woolley, B. Ray. *J. phys. chem. sol.*, **16**, 102, 1960.
371. V. A. Petrusevich, V. M. Sergeeva, *Fiz. tverd. tela*, **2**, 11, 2881, 1960.
372. S. I. Gadzhiev, K. A. Sharifov. In: *Problems of the Metallurgy and Physics of Semiconductors.* Izd. AN SSSR, 1961, p. 43.
373. V. P. Zhuze, A. I. Zaslavskii, V. A. Petrusevich, V. M. Sergeeva, I. A. Smirnov, A. I. Shelekh, *Proc. Intern. Conference on Semiconductor Physics.* Prague, 871, 1960.
374. I. Z. Fisher, *Fiz. tverd. tela*, **1**, 193, 1959.
375. G. Harbeke, G. Lautz. *Zs. Naturf.*, **13a**, 9, 771, 1958.
376. D. N. Nasledov, I. A. Feltyn'sh, *Fiz. tverd. tela*, **2**, 5, 823, 1960.
377. J. C. Woolley, B. A. Smith. *Proc. phys. soc.*, **72**, 867, 1958.
378. S. I. Radautsan, O. N. Derid, *Izv. Moldav. filiala AN SSSR*, **3** (69), 105, 1960.
379. H. Hahn, G. Frank, W. Klingler. *Zs. anorg. allg. Chem.*, **279**, 5–6, 271, 1955.
380. C. E. Olsen, P. M. Harris. *Phys. rev.*, **2**, 86, 659, 1952.
381. L. Suchow, P. H. Keck. *J. Amer. chem. soc.*, **75**, 518, 1953.
382. H. Hahn, G. Frank. *Zs. anorg. allg. Chem.*, **269**, 227, 1952.
383. N. A. Goryunova, V. A. Kotovich, V. A. Frank-Kemenetskii, *Dokl. Akad. Nauk SSSR*, **103**, 4, 639, 1955; *Zh. tekh. fiz.*, **25**, 14, 2419, 1955.
384. S. I. Radautsan, E. I. Gavrilitsa, *Izv. AN Moldav. SSR*, No. 10 (88), 95 (1961).
385. D. R. Mason, D. F. O'Kane. *Proc. Intern. Conference on semiconductor. Physics.* 1026 Prague. 1960.
386. D. R. Mason, J. S. Cook. *Bull. Amer. soc.*, **11**, 5, 1 (78), 1960.
387. Z. I. Ornatskaya. In: *Problems of the Metallurgy and Physics of Semiconductors*, M., Izd. AN SSSR, 1961, p. 145.
388. G. Busch, P. Junod, E. Mooser, H. Schade. *Halbleiter and Phosphore.* Braunschweig, S. 470, 1958.

389. L. S. Palatnik, Yu. F. Komnik, V. M. Koshkin, E. K. Belova, *Dokl. Akad. Nauk SSSR*, **137**, 1, 1961.
390. M. S. Mirgalovskaya, E. V. Skudnova, *Izv. AN SSSR, otd. tekh. nauk, Metallurgiya i toplivo*, **4**, 148, 1959.
391. S. I. Radautsan, I. A. Madan, R. A. Ivanova, *Izv. AN SSSR*, **10** (88), 98, 1961.
392. I. I. Kozhina, S. S. Tolkachev, A. S. Borshchevskii, N. A. Goryunova, *Vestnik Len. gos. univ.*, **4**, 122, 1962.
393. M. S. Mirgalocskaya, E. M. Komova. In: *Problems of the Metallurgy and Physics of Semiconductors.* Izd. AN SSSR, 1961, p. 138.
394. S. I. Radautsan, I. A. Madan, I. P. Molodyan, R. A. Ivanova, *Izv. Moldav. filiala AN SSSR*, **3** (69), 107, 1960.
395. N. A. Goryunova, S. I. Radautsan, *Zh. tekh. fiz.*, **28**, 9, 2917, 1958.
396. S. I. Radautsan, *Zh. nauch. khim.*, **4**, 1121, 1959.
397. N. A. Goryunova, S. I. Radautsan, *Dokl. Akad. Nauk, SSSR* **21**, 847, 1958.
398. N. A. Goryunova, S. I. Radautsan, G. A. Kiosse, *Fiz. tverd. tela*, **1**, 1858, 1959.
399. J. C. Woolley, C. M. Gillet, J. A. Evans. *J. chem. phys. sol.*, **16**, 138, 1960.
400. M. S. Mirgalovskaya, E. V. Skudnova, *Zh. nauch. khim.*, **4**, 1113. 1959.
401. D. N. Nasledov, I. A. Feltin'sh, *Fiz. tverd. tela*, **1**, 4, 565, 1959.
402. H. Hahn, D. Tiele. *Zs. anorg. allg. Chem.*, **303**, 147, 1960.
403. S. I. Radautsan, I. A. Madan. *Abstracts of the All-Union Conference on Semiconductors.* M., L., Izd. AN SSSR, 191, p. 56.
404. S. I. Radautsan, *Express Information.* Kishinev, Izd. GNTK Moldav. SSR, June, 1960.
405. N. A. Goryunova, S. I. Radautsan, V. I. Deryabina, *Fiz. tverd. tela*, **1**, 512, 1959.
406. D. N. Nasledov, M. P. Pronina, S. I. Radautsan, *Fiz. tverd. tela*, **2**, 1, 50, 1960.
407. L. I. Berger, S. I. Radautsan. In: *Problems of the Metallurgy and Physics of Semiconductors.* Izd. AN SSSR, 1961, p. 129.
408. J. C. Woolley, B. R. Pamplin, J. A. Evans. *J. phys. chem. sol.*, **19**, 1–2, 147, 1961.
409. S. I. Radautsan, Master's dissert. L., 1959.
410. B. R. Pamplin. *Nature.* Lond., **188**, 136, 1960.
411. G. A. Kiosse, G. I. Malinovskii, S. I. Radautsan, *Izv. Moldav. filiala AN SSSR*, **3** (69), 3, 1960.
412. S. I. Radautsan, I. P. Molodyan, *Izv. Moldav. filiala AN SSSR*, **3** (69), 37, 1960.
413. S. I. Radautsan, R. A. Ivanova, *Izv. AN Moldav. SSR*, No. 10 (88), 64, 1961.
414. E. Zintl, G. Brauer. *Zs. Elektr. angew. Phys. Chem.*, **41**, 102, 1935.
415. R. Brill. *Zs. Kristallogr. Mineral.*, **65**, 94, 1927.
416. M. V. Stackelberg. *Zs. phys. Chem.*, **27**, 53, 1934.

417. W. Frankenburger. *Zs. Elektrochem.*, **32**, 484, 1926.
418. G. Brauer, E. Zintl. *Zs. phys. Chem. Abt.* 8, **37**, 323, 1937.
419. B. Neuman, C. Kroger, H. Haebler. *Zs. anorg. allg. Chem.*, **204**, 81, 1932.
420. P. Lebeau. *Cr. Acad. sci. Paris*, **129**, 49, 1899.
421. P. Leneau. *Cr. Acad. sci. Paris*, **231**, 284, 134, 1902.
422. E. Zintl, G. Brauer. *Zs. Elektrochem., Bd.* **41**, 5, 297, 1935.
423. G. Grube, G. Vosskümler, H. Schlecht. *Zs. Electrochem.*, **40**, 270, 1934.
424. A. Joannos. *Compt. rend.* Paris, **119**, 559, 1894.
425. M. P. Morozova, G. A. Bol'shakova, I. L. Lukinykh, *Zh. obshch. khim.*, **29**, 3144, 1959.
426. G. Tammann. *Zs. phys. Chem.*, **3**, 446, 1889.
427. E. Zintl, J. Gonbeau, W. Dullenkopf. *Zs. phys. Chem.* Abt. A, **154**, 1, 1931.
428. C. Hugot. *Compt. rend. Paris*, **127**, 553, 1898.
429. C. Hugot. *Compt. rend. Paris*, **129**, 604, 1899.
430. E. Zintl, J. Goubeau, W. Dullenkopf. *Zs. Phys. Chem.* Abt. A, **154**, 32/33, 1931.
431. C. H. Mathewson. *Zs. anorg. allg. Chem.*, **50**, 192, 1906.
432. P. Lebeau. *Compt. rend. Paris*, **130**, 502, 1900.
433. C. H. Mathewson. *Zs. anorg. allg. Chem.*, **50**, 187, 1906.
434. A. Joannos. *Ann. chim. phys.*, **8** (7), 105, 1906.
435. A. Vournasos. *Zs. anorg. Chem.*, **81**, 366, 1913.
436. N. Parravano, *Garz. Chim. Ital.*, **45**, 485, 1915.
437. D. P. Smith. *Zs. anorg. allg. Chem.*, **56**, 126, 1908.
438. A. G. Vournasos. *Cr. Acad. Sci.* Paris, **152**, 714, 1911.
439. N. N. Zhuravlev, G. S. Zhdanov, R. N. Kuz'min, *Kristallografiya*, **5**, 1, 134, 1960.
440. N. N. Zhuravlev, T. A. Mingazin, T. S. Zhdanov, *Zh. eksper. tekh. fiz.*, **34**, 820, 1958.
441. N. S. Zaitsev, *Zh. tekh. fiz.*, **9**, 661, 1939.
442. N. D. Mrgulis, B. I. Dyatlovitskaya, *Zh. tekh. fiz.*, **10**, 657, 1940.
443. N. S. Khlebnikov, *Zh. tekh. fiz.*, **17**, 333, 1947.
444. S. Yu. Luk'yanov, I. M. Mazover, *Zh. eksper. teoret. fiz.*, **9**, 1459, 1939.
445. N. S. Khlebnikov, N. S. Zaitsev, *Zh. tekh. fiz.*, **9**, 44, 1939.
446. A. Sommer. *Proc. Phys. soc.*, **55**, 145, 1943.
447. P. G. Borzyak, *Fizich. zap. AN SSSR*, **9**, 173, No. 2, 1941.
448. R. Juza, H. Hahn. *Zs. anorg. Chem.*, **241**, 177, 1939.
449. R. Juza, K. Bär, *Zs. anorg. Chem.*, **283**, 230, 1956.
450. H. Haraldsen. *Zs. anorg. allg. Chem.*, **240**, 337, 1939.
451. O. Kubaschewski, E. Evans, *Thermochemistry in Metallurgy*. (Translation into Russian). M., IL, 1954.
452. W. Hofmann. *Zs. Metallkunde*, **33**, 61, 1941.
453. R. Vogel, R. Dobbener, O. Strathmann. *Zs. Metallkunde*, **50**, 3, 130, 1959.

454. G. A. Eade, W. Hume-Rothery. *Zs. Metallkunde,* **50**, 3, 123, 1959.
455. H. Reinhold, R. Schmitt. *Zs. phys. Chem.,* **44**, 75, 1939.
456. C. Tubandt, H. Reinholdt. *Zs. Electrochem.* **37**, 589, (1931).
457. P. Rahlfs. *Zs. phys. Chem.* **31**, 157, 1935.
458. A. Boettcher, G. Haase, H. Treupel. *Zs. angew. Phys.,* **7**, 478, 1955.
459. F. Klaiber. *Annall. Phys.* **5**, 3, 229, 1929.
460. C. Wagner. *Zs. phys. Chem.* **23**, 469, 1933.
461. C. Tubandt, H. Reinhold. *Zs. phys. Chem.,* **24**, 22, 1934.
462. F. C. Kracek. *Trans. Amer. Geophys. Union,* **27**, 364, 1946.
463. H. Reinhold, H. Möhring. *Zs. phys. Chem.,* **28**, 178, 1935.
464. P. Junod. *Helv. Phys. acta,* **32**, 6/7, 567, 1959.
465. G. Busch, P. Junod. *XVII Intern. Kongr. München,* 140, 1959.
466. J. Coun, R. C. Taylor. *J. Electrochem. Soc.,* **107**, 12, 977, 1960.
468. J. Appel. *Zs. Naturforsch,* **10ᵃ**, 7, 530, 1954.
469. J. Appel, G. Lautz. *Physica,* **20**, II, Novemb. 1954.
470. H. Reinhold, H. Seidel. *Zs. phys. Chem.,* **38**, 245, 1937-1938.
471. H. Reinhold, H. Möhring. *Zs. phys. Chem.,* **38**, 221, 1937-1938.
472. H. Nowotnu. *Metallforsch.,* **1**, 40, 1946.
473. G. Busch, U. Winkler. *Physica,* **20**, 1067, 1954.
474. U. Winkler. *Helv. Phys. Acta,* **28**, 633, 1955.
475. L. Wöhler, O. Schliephake. *Zs. anorg. allg. Chem.,* **151**, 11-20, 1926.
476. R. Vogel. *Zs. anorg. allg. Chem.,* **61**, 46-53, 1909.
477. E. Zintl, H. Kaiser. *Zs. anorg. allg. Chem.,* **211**, 125-31, 1933.
478. R. F. Blunt, HPR Frederiks, W. R. Hosler. *Bull. Amer. phys. soc.,* **30**, 2, 7, 1955.
479. B. I. Boltaks, *Dokl. Akad. SSSR,* **14**, 487, 1949.
480. B. I. Boltaks, *Zh. tekh. fiz.,* **20**, 180, 1950.
481. N. S. Kurnakov, N. I. Stepanov, *Zh. russ. fiz. khim. obshch.,* **37**, 568, 1905.
482. W. Klemm, H. Westlinning. *Zs. anorg. Chem.,* **245**, 365, 1941.
483. W. D. Robertson, H. H. Uhlig. *J. Elektrochem. soc.,* **96**, 27, 1949.
484. W. D. Robertson, H. H. Uhlig. *Metals technology,* 2468, **15**, 5-8, 1948.
485. W. Hume-Rothery. *J. inst. met., London,* **38**, 127-31, 1927.
486. A. Knappwost. *Zs. Electrochem.,* **56**, 594, 1952.
487. B. Friant. *J. Amer. chem. soc.,* **48**, 1906, 1926.
488. A. Sacklowski. *Ann. phys.,* **77**, 241, 1925.
489. N. J. Stepanow. *Zs. anorg. allg. Chem.,* **60**, 209!229, 1908.
490. A. Saklowski. *Ann. phys.,* **77**, 264, 1925.
491. G. Busch, P. Junod, U. Katz, U. Winkler. *Helv. phys. acta,* **27**, 193, 1954.
492. P. Eckerlin, E. Leicht, E. Wölfel. *Zs. anorg. allg. Chem.,* **280**, 321, 1955.
493. O. Schliephake. *Diss.* Darmstadt 1926.
494. L. Wöhler, O. Schliephake. *Zs. anorg. allg. Chem.,* **151**, 1/11, 1926.
495. P. Eckerlin, E. Leicht, E. Wölfel. *Zs. anorg. allg. Chem.* **307**, 3-4, 143, 1961.

496. W. Hume-Rothery. *J. inst. met.*, London, **35**, 319/35, 1926.
497. N. Baar. *Zs. anorg. allg. Chem.*, **70**, 372/77, 1911.
498. M. Stackelberg. *Zs. phys. Chem.*, **27**, 50, 1934.
499. H. A. Sloman. *J. inst. met.*, London, **49**, 370, 1932.
500. M. V. Stackelberg. *Zs. Electrochem.*, **37**, 542/45, 1931.
501. M. Stackelberg, E. Schnorrenberg. *Zs. phys. Chem.*, **27**, 37/49, 1934.
502. G. Oesterheld. *Zs. anorg. allg. Chem.*, **97**, 1, 1916.
503. M. Lebeau. *Cr. Acad. sci.* Paris, **121**, 496, 1895.
504. C. Fichter, E. Brunner. *Zs. anorg. allg. Chem.*, **93**, 91, 1915.
505. M. Stackelberg, R. Paulus. *Zs. phys. Chem.*, **22**, 305, 1933.
506. H. H. Franck, H. Füldner. *Zs. anorg. allg. Chem.*, **204**, 97, 1932.
507. B. Neumann, C. Vrager, H. Haebler. *Zs. anorg. allg. Chem.*, **207**, 138/41, 1932.
508. L. I. Mirkin, *Handbook of X-ray Diffraction Analysis of Polycrystals.* M., Fizmatgiz, 1961.
509. C. Matignon. *Compt. rend.* Paris, **154**, 1351/53, 1912.
510. T. P. Blunt. *J. chem. soc.*, **3**, 106/8, 1865.
511. J. Parkinson. *J. chem. soc.*, **20**, 117, 1867.
512. H. Gautier. *Compt. rend.* Paris, **128**, 1167/69, 1899.
513. G. Natta, L. Passerini. *Gazz.*, **58**, 541, 655, 1928.
514. Parkinson. *J. chem. soc.*, **20**, 127, 309, 1867.
515. V. P. Zhuze, I. V. Mochan, S. M. Ryvkin, *Zh. tekh. fiz.*, **18**, 1494, 1948.
516. B. I. Boltaks, V. P. Zhuze, *Zh. tekh. fiz.*, **18**, 12, 1459, 1948.
517. T. S. Moss. *Proc. phys. soc.*, **63**, 982, 1950.
518. P. G. Borzyak, L. S. Miroshnichenko, R. D. Fedorovich, *Fiz. tverd. tela*, **3**, 6, 1778, 1961.
519. G. Grube, R. Bornhak. *Zs. Elektrochem.*, **40**, 140/42, 1934.
520. N. I. Stepanov, *Zh. russ. fiz. khim. obshch.*, **37**, 1285, 1905.
521. G. Grube. *Zs. anorg. allg. Chem.*, **49**, 83, 87, 1906.
522. A. Antropoff, E. Germann. *Zs. phys. Chem.*, **137**, 209, 1928.
523. A. Antropff, F. Falk. *Zs. anorg. allg. Chem.*, **187**, 405, 1930.
524. J. Rieber. *Diss.* Hannover, 24, 1930.
525. A. Grantz, G. Rederer. *Bull. soc. chim.*, **3**, 35, 510, 1906.
526. A. Jaboin. *Compt. rend.* Paris, **129**, 763, 1900.
527. *Weldgin engineer*, **43**, 4, 97, 1958.
528. A. Sieverts, W. Krumbhaar. *Ber. dtsch. chem. Ges.*, **43**, 894 (1910).
529. R. Juza, H. Hahn. *Zs. anorg. allg. Chem.*, **244**, 125, 1940.
530. R. Juza, H. Hahn. *Zs. anorg. allg. Chem.*, **239**, 273, 1938.
531. A. J. Rosenberg, T. C. Harman. *J. appl. phys.*, **30**, 1621, 1959.
532. M. V. Stackelberg, R. Paulus. *Zs. phys. Chem.*, Abt. **28**, 427, 1935.
533. P. Jolibois. *Compt. rend.* Paris. **147**, 801 (1908).
534. S. A. Shchukarev, G. Grossman, M. P. Morozova, *Zh. org. khim.*, **25**, 633, 1955.
535. M. Kavelis, *Uch. zap. Vil'nyusskogo univ.*, Khimiya, **28**, 111, 1959.
536. G. V. Samsonov, L. L. Vereikina, *Phosphides.* Kiev, Izd. AN USSR, 1961.

537. G. A. Silvey. *J. appl. phys.*, **29**, 2, 226, 1958.
538. H. Cole, F. W. Chambers, H. M. Dunn. *Acta cryst.*, **9**, 685, 1956.
539. W. J. Turner, A. S. Eischler, W. F. Reese. *Phys. rev.*, **121**, 3, 759, 1961.
540. W. Heike. *Zs. anorg. allg. Chem.*, **118**, 264–268, 1921.
541. A. Descamps. *Compt. rend.* Paris, **86**, 1066, 1878.
542. S. F. Zemczuzny. *Zs. anorg. allg. Chem.*, **49**, 384, 1906.
543. Z. Regnault. *Compt. rend.* Paris, **76**, 23, 1878.
544. W. Oppenheim. *Ber. drsch. chem. Ges.*, **5**, 979, 1872.
545. S. A. Shchukarev, M. P. Morozova, M. M. Bortnikova, *Zh. org. khim.*, **28**, 3289, 1958.
546. T. S. Moss. *Proc. phys. soc.*, **63**, 167, 1950.
547. W. Zdanowicz. *Proc. Intern. conference on semiconductor physics*, 1095. Prague, 1960.
548. S. A. Shchukarev, M. P. Morozova, M. M. Bortnikova, *Zh. org. khim.*, **28**, 12, 3289, 1958.
549. W. Treitschke. *Zs. anorg. Chem.*, **50**, 217, 1906.
550. R. Juza, W. Uphoff, W. Gieren. *Zs. anorg. allg. Chem.*, **292**, 71, 1957.
551. H. Nowotny, K. Bachmayer. *Monatshefte Chem.*, **81**, 488, 1950.
552. H. Nowotny, K. Bachmeyer. *Monatshefte Chem.*, **80**, 734, 1949.
553. F. Laves, J. D. Ans, E. Lax. *Taschenbuch f. Chemiker und Phys.*, 2. Aufl., 170, 1949.
554. R. Junod, E. Mooser, H. Schade. *Helv. phys. acta*, **29**, 193, 1956.
555. F. Laves, D. Ans. *Taschenbuch f. Chemiker und Phts.* 1. Aufl., 1943.
556. H. Nowotny, B. Glatze. *Monatshefte Chem.*, **83**, 237, 1952.
557. H. Nowotny. *Zs. Metallkunde*, **34**, 237, 1942.
558. H. Nowotny, B. Glatze. *Monatshefte Chem.*, **82**, 720, 1951.
559. H. Nowotny, W. Sibert. *Zs. Metallkunde*, **33**, 391, 1941.
560. L. Dupare, P. Wenger, C. Urfer. *Helv. chim. acta*, **13**, 657, 1930.
561. R. Juza, W. Schulz. *Zs. anorg. allg. Chem.*, **269**, 1–2, I, 1952.
562. W. Frankenburger, L. Andrussov, E. Durr. *Zs. Elektrochem.*, **34**, 632, 1928.
563. R. Juza, H. Hermann. *Zs. anorg. Chem.*, **273**, 48, 1953.
564. S. A. Shchukarev, *Zh. org. khim.*, **24**, 4, 582, 1954.
565. E. V. Biron, *Zh. russk. fiz. khim. obshch.*, **47**, 964, 1915.
566. H. Weiss. *Ann. Phys.*, **4**, 121, 1959.
567. S. Stil'bans. In: *Semiconductors in Science and Technology.* M., L., Izd. AN SSSR, **1**, 86, 1957.
568. B. F. Ormont, *Zh. nauch. khim.*, 3, 6, 1281, 1958.
569. S. M. Ariya, *Zh. org. khim.*, **27**, 6, 1405, 1957.
570. B. F. Ormont, N. A. Goryunova, N. N. Ageeva, N. N. Fedorova, *Izv. AN SSSR, ser. fiz.*, **21**, 1, 1957.
571. L. R. Weisberg, J. Blanc. *Proc. Intern. Conference on semiconductor physics* 940, Prague 1960, publ. in 1961.
572. S. Z. Roginskii, *Zh. vsesoyuz. khim. obshch.*, im. D. I. Mendeleeva, **5**, 5, 482, 1960.
573. O. V. Krylov, S. Z. Roginskii, *Izv. org. khim. nauk*, 1, 17, 1959.

574. O. V. Krylov, E. A. Fokina, *Zh. fiz. khim.*, **35**, 3, 651, 1961.
575. S. Z. Roginskii. In: *Problems of Kinetics and Catalysis*, **10**, 5, 1960.
576. A. F. Ioffe, S. V. Airapetyants, A. V. Ioffe, N. V. Kolomoets, L. S. Stil'bans, *Dokl. Akad. Nauk SSSR*, **106**, 981, 1956.
577. G. A. Geffrey, G. S. Perry, R. L. Mozzi. *J. chem. phys.*, **25**, 1024, 1956.
578. E. Burstein, P. H. Eggli. *Advances in Electronics*, **7**, 24, 1955.
579. R. F. Potter. *J. phys. chem. sol.*, **3**, 223, 1957.
580. B. R. Bäuerlein. *Zs. Naturf.*, **14a**, 1069, 1959.
581. H. Welker. *Zs. Naturf.*, **7a**, 744, 1952.
582. A. I. Gubanov, *Zh. tekh. fiz.*, **26**, 10, 2170, 1956.
583. V. P. Zhuze, *Zh. tekh. fiz.*, **25**, 12, 2079, 1955.
584. B. F. Ormont, *Dokl. Akad. Nauk SSSR*, 124, 129, 1959.
585. C. H. L. Goodman. *Brit. j. electronics*, **1**, 2, 115, 1955.
586. O. G. Folberth. *Zs. Naturf.*, **13a**, 856, 1958.
587. O. G. Folberth. *Zs. Naturf.*, **15a**, 425, 1960.
588. W. P. Pearson. *Canad. j. chem.*, **37**, 1191, 1959.
589. P. Manga. *J. phys. chem. sol.*, **20**, 3–4, 268, 1961.
590. E. P. Stambauch, Y. I. Genco, R. C. Himes. *I. Electroch. soc.*, **107**, 3. 1960.
591. A. V. Petrova, E. L. Strum, *Fiz. tverd. tela*, **4**, 6, 1442, 1962.
592. L. S. Palatnik *et al.*, *Fiz. tverd. tela*, **4**, 6, 1430, 1962.
593. Ya. K. Syrkin, *Usp. khimii*, **31**, 4, 397, 1962.

General Index

A^1B^7, 2, 14, 123
A^2B^6, 2, 14, 85, 117, 121, 138, 160
A^3B^5, 1, 2, 14, 83-89, 148, 160, 195-196, 199-200, 202*ff*, 215
A^3B^5-A^2B^6, 148
$A^1B^3C_2^6$, 12, 140*ff*
$A^2B^4C_2^5$, 12, 142-143, 215
$A_2^1B_2^4C_3^5$, 12, 144
$A_2^1B^4C_3^6$, 12, 145
$A_3^1B^5C_4^6$, 12, 146
A_2B_3, 12, 42, 153-156
$A_2^1B^2C_4^7$, 158
$A_2^2B_2^3C_4^6$, 159, 163
$A_2^2B^4C_4^6$, 166
$A_3^5B^5C_3^6$, 166
$A^2B^4C_3^6$, 176
$A^3B^5C_4^6$, 176
Absorption edge, 62
AHRENS, 8-9
Alloying, 61
Aluminium, 6
Antifluorite structure, 47
Apatite, 13
Atomic number and semiconducting properties, 193-198

Band gap, *see* forbidden energy gap
Band structure of Ge and Si, 204
BERGMAN, 95
Beryllium, 6
BESSON, 92
Binary compounds, 1, 3, 12-15, 201, *and see* under specific compounds

Bonds, 9
  covalent, 8-9, 40, 200, 203-204
  homopolar, 196
  ionic, 6-9, 40, 204-208
  ionic-covalent, 201
  metallic, 5-9, 200*ff*
  valence, 7
Boron, 6
Borazone, *see* BN
BRODER, 89

Caesium chloride structure, 5
Carrier mobility, *see* mobility
Catalytic properties of diamond-like semiconductors, 203
Chalcogenides, 120*ff*, 162, 154-155, 204
Chalcopyrite, 141
  structure, 140-142, 149-150, 160, 163
Cleavage, 59, 83
Close packing, 6, 55, 84
Cohesive forces, 196, 206
Conduction
  impurity, 61-62
  intrinsic, 60, 62, 203
Co-ordination
  number, 1
Criteria for new semiconductors, 56
Crystallochemical series, 2, 3, 5, 8, 42, 54
Cuprous halides, 123
CZOCHRALSKI, 88, 100, 105

Debye temperature, 199

Defect tetrahedral phases,	46, 152*ff*	Fourfold co-ordination,	3, 5
Defect semiconductors,	12, 42, 145	Frenkel defects,	206, 211

Diamond
   electrical and physical properties, 64-65
   structure, 2, 13

Dicationic systems
   ternary, 15*ff*, 149
   quaternary, 19
   quinary, 28

Dielectric constant, 200
Dissociation energy, 201, 207
Distribution coefficients of
   impurities, 61
   in Ge and Si, 69
   in InAs, 169
   in InSb, 116
DO_3 structure, 47

EFFER, 110
Electron
   affinity constants, 7, 9, 195
      table of, for elements, 10
      specific, 9-12
   configuration from group theory, 6
   negativity, 6, 11
Electrons
   per atom, 44*ff*, 54*ff*
   density, 8
   d, 6, 56, 193
   dsp, 6
   f, 193
   sp, 56, 64
   valence, 8, 9, 13
Electronic
   conduction, 61, 86, 196
   polarisation, 200
Enargite, 146
Etchants for Ge and Si, 67, 74
Ettingshausen-Nernst coefficients, 112, 131
Eutectic point, 202

Famatinite, 146
FERSMAN, 13
FOLBERTH, 108, 211
Forbidden energy gap, 7, 60, 85, 200, 207-209

Germanium, 58, 86
   comparison with silicon, 67
   effects in stabilising grey tin, 82
   preparation, 75
   properties
      chemical, 74
      electrical, 67, 76
      optical, 76
      physical, 73
   and silicon alloys, *see* silicon-germanium alloys
GLASOV, 95
GOLDSCHMIDT, 1, 2, 6, 12, 56, 94, 196, 204, 206, 212, 216
GOODMAN, 13, 210
GRIMM, 12
GUERTLER, 95
GYBANOV, 208

HAHN, 153, 164
Hall
   effect, 86
   transducers, 86, 114
Hardness, 59, 196, 206
   and atomic distance, 196
   effect of illumination, 70
   micro-, 59
HARDY, 212
HERMAN, 204
Hetero substitution, *see* transverse hetero substitution
Hole conductivity, 61, 85
Homopolar bonds, *see* bonds

Impurity conduction, *see* conduction
Intrinsic conduction, *see* conduction
Infra-red detectors, 87
IOFFE, A. F., 127
IOFFE, A. V., 59, 82, 127, 204
Ionic
   cores, 8
   covalent bonds, *see* bonds
   radius, 8
Ionisation
   energies, 193

group,	195
potential,	6-10
Isoelectronic series,	2-3, 41, 204-208
Keck,	89
Kimball,	6
Klingler,	153
Knoop	
hardness,	212
pyramid,	59
Kurnakov,	59
Lattice spacing	
of binary compounds,	2
in ionic series,	205
in isoelectronic series,	2, 205
Lead,	200
Lickhtman,	87
Lifetime	
of current carriers,	63-64, 85
Magnesium,	6
Magnus' Law,	5-6
Maslennikov,	87
Mendeleev,	1, 72, 193
Mercury chalcogenides, see chalcogenides	
Metallic bonds, see bonds	
Microhardness, see hardness	
Mobility 60, 86, 117, 201, 208-209, 214 and see under specific compounds	
Molecular structures,	45
Monocationic	
ternary systems,	15ff, 52, 146
quaternary systems,	19
quinary systems,	28
Monocrystals	
preparation,	87
Moissan,	92
Mooser,	6-7, 11, 56
Morris,	8
Morphotropic transitions,	5
Nekraskov,	7-8
Octahedral structures,	11, 55

Optical	
absorption,	86
absorption edge, see absorption edge	
properties, 62, 200 and see under specific compounds	
Pauling,	212
Pearson,	6-7, 11, 56, 211
Periodic table	
and crystal structure,	4
Phase diagrams, 89, 154 and see under specific compounds	
Phases of variable composition,	20, 41, 52
Photocells,	85
Photoelectric effect,	62, 153, 200, 214
Polarisability,	8
Polarisation,	6
P-n junctions,	61, 105
Pseudosections	
binary,	19
ternary,	21
Quantum number,	6, 8
Quasi-binary systems,	21
Quaternary systems,	13, 19ff, 53, 146ff, 202
based on A^2B^6,	139, 151
based on A^3B^5,	147
based on ternary phases,	151
Quinary systems,	28ff, 150ff
Radius	
of cation,	8
ionic,	9
Refractive index,	200
Regel,	204
Rock salt structure,	6
Roginskii,	203
Roozeboom,	129, 132, 150
Ruherwein,	92
Secondary periodicity,	193
Semiconduction	
and chemical properties,	60
and optical properties,	62
and thermodynamics,	59

Sextuple systems, 152
SHCHUKAREV, 193
Short range order, 55, 58, 84
Silicon-germanium alloys, 77-80
Silver halides, 124
Solar batteries, 86, 102
Solid solutions, 203
SOMMERFELD, 12
Spinel structure, 159
Sphalerite structure, 2-6, 12, 44, 54, 163
STOHR and KLEMM, 124
Stoichemetric composition, 84

Ternary systems, 15*ff*, 202
  based on A^1B^7, 139
  based on A^2B^6, 138
  based on A^3B^5, 125*ff*
  based on binary defect structures, 46*ff*
Tetrahedral phases
  binary, 12-13, 41
  complex, 12-13, 40-43
  defect, 41-46, 53, 152-157
  excess, 42, 46, 176
  rules for, 14
  ternary, 17
  structures, 3-6, 9-11, 13
Tetracationic quinary systems, 28
Tin
  grey, 58
  effect of germanium, 81
  preparation and properties, 81-82
  white, 80*ff*

Thermal conductivity, 199
  in defect structures, 155-157
Thermal expansion coefficient, 207
Thermoelectric
  generator, 86
  properties, 58, 86, 192, 216
Thin films, 146
  anomalous crystal structure, 4
Thiogallate structures, 162
THOMAS, 212
Transistors, 214
Transport processes, 89
Tricationic systems
  quaternary, 19
  quinary, 28

URAZOV, 95

VERNASKII, 13
Valency, 26, 42, 54-55
Vapour pressure
  of elements, 73
Vegard's Law, 127, 135, 149

WELKER, 207-209
WILLIAMS, 92
WINKLER, 72
WOLFF, 89
Wurtzite structure, 2-6

ZHUZE, 209
Zinc blende structure, *see* sphalerite
Zone levelling, 75
Zone refining, 62*ff*, 85

Formula Index

Elements are listed in main index. Compounds in this index are in alphabetical order of initial symbol. Further symbols in each formula are in the order of the groups of the period table.

AlN,	93	CdX_2Y_4 (X = Al, Ga, In;	
AlP,	94	Y = S, Se, Te),	160, 165
–ZnS,	148	$CdIn_2X_4$ (X = Se, Te),	47
AlAs,	94, 96	$CdSnX_2$ (X = P, As),	143, 145
–InAs,	126	$CdSnSb_2$,	40
AlSb,	95-97	CdX (X = S, Se),	120
–Al_2Te_3,	167	CdTe,	82
–CdTe,	148	–$XInTe_2$ (X = Cu, Ag),	151
–GaSb,	105, 127-128	Cd_2GeX_4 (X = S, Se),	166
–InSb,	129-131	Cd_3X_2 (X = N, P, Sb),	184, 186
–Al_2Te_3,	167	Cd_3As_2,	176, 217
Al_2X_3 (X = S, Se, Te),	155	CoMnSb,	188
Ag X As (X = Mg, Zn),	188	Cu X Y (X = Mg, Cd;	
$AgGe_2P_3$,	145	Y = As, Sb, Bi),	188
$AgFeTe_2$,	142	$CuInSe_2$,	141
Ag_2HgI_4,	47	$CuGeSe_3$,	41
Ag_2 X (X = S, Se, Te),	180	$CuGe_2X_3$ (X = S, Se, Te),	
Ag_3 X (X = N, P, As, Sb),	179		41, 144, 145
Ag_3S X (X = I, Br),	47, 188	$CuAsX_2$ (X = S, Se),	142
Au_3P,	179	$CuGe_2As_3 - Cu_2GeSe_3$,	151
		Cu_2HgI_4,	47
BX (X = N, P, As, Sb),	90-93	Cu_2XY_3 (X = Ge, Si, Sn;	
$BeSiN_2$,	142	Y = S, Se, Te),	145
BeX (X = O, S, Se, Te),	118	Cu_2X (X = S, Se, Te),	180
Be_2C,	183	Cu_3X (X = N, P, As, Sb),	179
Be_3X_2 (X = N, P),	184		
Ba_3X_2 (X = N, P),	186	GaN,	98
		GaP,	98-99, 215
Ca_3X (X = N, P),	184	–GaAs,	134$f\!f$

–InP,	125
–Si – ZnSiP$_2$,	150
GaAs,	99–103
–GaSb,	134
–InAs,	126, 128
–ZnTe,	148
–ZnGeAs$_2$,	148
GaSb,	103–105
–InSb,	105, 132–134
GaBr$_3$,	43
Ga$_2$ZnX$_4$ (X = Se, Te),	163, 164
Ga$_2$X$_3$ (X = S, Se, Te),	44, 153, 157
–GaP,	167, 168
–GaAs,	167–170
–GaSb,	167, 171
Ga$_2$Te$_3$ XTe (X = Zn, Cd, Hg),	163
Ga$_3$As$_4$,	47
Ga$_4$Ge$_3$,	47
GeS$_2$,	44
Ge$_3$N$_4$,	47
Ge$_3$P$_4$,	145
HgX$_2$Y$_4$ (X = Al, Ga, In;	
Y = S, Se, Te),	160, 165
HgX (X = S, Se, Te),	86, 119–120
HgI$_2$,	45
–X$_2$I$_2$ (X = Cu, Ag),	158
Hg$_2$GeX$_4$ (X = S, Se),	166
Hg$_3$P$_2$,	186
InN,	105–106
InP,	106–108
–CdX (X = S, Se, Te),	148
–InAs,	135–136
–In$_2$X$_3$ (X = S, Se, Te),	171
InAs,	86, 109–112
–ZnXAs$_2$ (X = Ge, Sn),	148
–CdSnAs$_2$,	149
–HgTe,	148
–InSb,	136–137
–In$_2$X$_3$ (X = S, Se, Te),	171
InSb,	82, 86, 105, 112–116
–Ag$_3$SbTe$_4$,	150
–X SnSb$_2$ (X = Zn, Cd),	148
–CdTe,	148
–In$_2$X$_3$ (X = S, Se, Te),	175
In$_2$Se$_3$	
–HgSe,	165
In$_2$Te$_3$,	153ff
K$_3$X (X = As, Sb, Bi),	178
LiXY (X = Mg, Zn;	
Y = N, P, As, Sb, Bi),	186
Li$_2$MgSn,	46
Li$_3$XY (X = Al, Ga;	
Y = N, A, As),	188
Li$_3$X (X = N, P, As, Bi),	42–45, 177
Li$_3$FeN$_2$,	190
Li$_5$XY$_3$ (X = Ge, Si, Ti),	190
MgGeP$_2$,	142
Mg$_2$X (X = S, Se),	4
Mg$_3$X$_2$ (X = N, P, As, Sb, Bi),	184
MnX (X = S, Se),	4
NaXAs (X = Mg, Zn),	188
Na$_3$X (X = P, As, Sb, Bi),	44, 45, 177, 178
NiMgX (X = Sb, Bi),	188
Rb$_3$X$_2$ (X = Sb, Bi),	179
Sr$_3$X$_2$ (X = N, P),	184
SiC,	58, 70–72, 215
SiS$_2$,	53
X$_2$Y$_3$ (X = B, N, Ga, In, Tl;	
Y = O, S, Se, Te),	154, 155
ZnXY$_4$ (X = Al, Ga, In;	
Y = Se, Te),	47, 160, 163, 165
ZnGeX$_3$ (X = S, Se),	52
ZnGeAs$_2$,	41
ZnO,	119
ZnS,	120
–XGaS$_2$ (X = Cu, Ag),	151
–GeX$_2$ (X = S, Se),	176
Zn$_2$GeX$_4$ (X = S, Se),	52, 162, 166
Zn$_3$X$_2$ (X = N, P, As),	45, 184